Augusta's

WGAC RADIO

DATE DUE

Demco, Inc. 38-293

MAY 3 0 2012

Augusta's
WGAC RADIO

THE VOICE OF THE GARDEN CITY
FOR SEVENTY YEARS

Edited by
Debra Reddin van Tuyll
and Scott Hudson

Charleston | London

THE
History
PRESS

Published by The History Press
Charleston, SC 29403
www.historypress.net

Cover: Front cover image from 123royaltyfree.com. Back cover image of WGAC studios
courtesy of John Diskes; WGAC ad courtesy of the *Augusta Chronicle*.

First published 2012

Manufactured in the United States

ISBN 978.1.60949.339.4

Library of Congress CIP data applied for.

Notice: The information in this book is true and complete to the best of our knowledge. It is
offered without guarantee on the part of the author or The History Press. The author and
The History Press disclaim all liability in connection with the use of this book.

CONTENTS

CONTENTS

PREFACE

O ver-the-air broadcast radio is a survivor. Television did not kill it off. Neither did the Internet, convergence or satellite radio. Across America, historic radio stations still broadcast to large audiences daily, but of those stations only a few can claim the iconic fame of Augusta's WGAC radio.

WGAC has inspired dreams and launched the careers of internationally known musicians, presidential candidates and business tycoons. In the early 1940s, *Life* magazine proclaimed the little five-thousand-watt AM station the "mirror of the industry." Little did its audience know at the time what an impact those five thousand watts would have on the state and the nation.

What started as the dream of a North Carolina farm boy nearly ninety years ago stands today as a media institution that is the envy of many. In a time when traditional media are losing market share to new digital media, WGAC continues to draw listeners and advertisers and have extraordinary influence within its community and state.

WGAC has not always been the number one radio station in the Augusta market, although for the last decade, the Georgia Broadcasters Association has named it the top radio station in the state among stations broadcasting in mid-sized markets. The station has had as many troughs in its history as peaks, but even through those downtimes, WGAC has kept its focus on serving its community, and it has earned a well-deserved reputation for excellence in its more than seventy years.

When WGAC first went on the air in 1940, commercial radio was less than twenty years old. The station, like radio itself, was truly an invention of the modern age. Guglielmo Marconi and others began their experiments with electromagnetic radiation, commonly known as radio waves, in the waning days of the nineteenth century. In 1900, he obtained his famous Patent No. 7777 for "tuned or syntonic telegraphy," and as the twentieth century dawned, he transmitted the first wireless broadcast from Cornwall, England, to St. John's, Nova Scotia.[1]

Marconi, Fessenden, Faraday and the other early radio pioneers were working from the idea that if one could transmit a Morse code signal over the air, it must also be possible to transmit a voice using simple telephone technology. As a practical matter, no one at the time, perhaps not even the scientists and inventors working with the fledgling radio technologies, saw much of a future in electromagnetic radiation transmission. Why spend the money on "wireless" when the infrastructure for wired telegraphy was already standard across the world? Wireless also had its limitations since it took enormous power to transmit a message over long distances, and relay stations were needed to receive and retransmit the information. Further, messages sent over wireless enjoyed no privacy, as anyone with a receiver could decode the message being sent.

Hobbyists and businesses were the first to find applications for radio. Within a decade of Marconi's first broadcast, nearly one thousand hobbyists had constructed radio sets that they used to connect with other amateurs. Their communications were almost exclusively by Morse code. The same was true for the shipping industry, the first business to find a practical application for radio, thanks to legislation passed by Congress.[2]

The Wireless Ship Act of 1910 required oceangoing vessels be equipped with a "wireless shack" if they carried fifty or more people between ports two hundred or more miles apart. Coded messages from ship to ship and from ship to shore meant that mariners could be warned of potential dangers, coordinate arrival times and come to one another's aid in the event of a problem at sea. Yet over that decade, wireless was used less for the transmission of serious information and more for wealthy transatlantic cruise passengers sending electronic postcards to friends on other ships and family back on land. Radio was an unregulated novelty. A massive tragedy at sea would change all that.[3]

On April 10, 1912, the RMS *Titanic* steamed out of Belfast Harbor, headed for New York City. Five days later, the world's largest, and allegedly unsinkable, ship hit an iceberg and sank. No one at the time would have dreamed that the sinking of that ship would open the way for the commercialization of a new communications technology that would have the power to bring voices from thousands of miles away right into American living rooms—as well as spawn thousands of urban legends, including that the *Titanic* was the first ship ever to send out an SOS signal. That honor actually goes to an American ship, the SS *Arapahoe*, which in 1909 sent the first distress signal using those letters. It was the *Titanic*, though, that made those letters famous and led to the rapid development of the medium of radio.

Radio developed in the wake of the *Titanic* disaster for two simple reasons. First, the disaster clearly demonstrated that radio had the power to save lives. If there had been no radio distress calls, everyone aboard that ship would have perished. Without radio, those 710 passengers and crew members lucky enough to cheat death in lifeboats would have died long before anyone would have known that the ship was missing. The famous Molly Brown would not have been so unsinkable were it not for radio operators Jack Phillips and Harold Bride begging for assistance from their tiny wireless room. It also became evident early on in the official RMS *Titanic* inquiry that most of the people aboard *Titanic* could have been saved if the radio operator aboard the SS *Californian*, which floated a mere ten miles away, had been at his post rather than asleep in his bunk.[4]

Because ship travel at the time was the only way to move between the continents, and as more people than ever were using that means of transport, it became obvious to those crafting maritime law that radio was every bit as much a safety device as a lifeboat. However, to the general public across the world, radio suddenly became more than a mere safety device; it became a way to transmit information from anywhere to everyone with an immediacy that newspapers could never match. Radio, in effect, set in motion what we now refer to as the twenty-four-hour news cycle.

A decade after the *Titanic* disaster, radio had been transformed from a novelty into the first true mass communication medium when the first commercially licensed radio station, KDKA of Pittsburgh, Pennsylvania, signed on the air in 1920 and began broadcasting to the general public—or at least to those who owned a radio set. Radio became a great tool for unification by delivering vast amounts of information to everyone at once. It would soon become an entertainment medium as well. With radio, people living and farming in rural areas suddenly found the almanac a helpful tool but no longer a necessity, for they could get farm reports from the wireless. Radio storm warnings meant that people could be better prepared for bad weather, and commercially sponsored entertainment shows made everyday household chores seem less chore-like.

Radio touched every segment of society and altered culture in a way like nothing had before. The illiterate could learn about important events even without being able to read. Politics in America changed as people could hear a candidate for office explain his platform with his own words using his own voice. Voters no longer had to travel away from home to hear a political speech because they could listen in their living room. Even the American presidency changed. Citizens could listen to the chief executive's voice. He became a real person, not simply a grainy picture of some man who lived in Washington, D.C.

Politicians learned to use radio to draw the country together during hard times. Franklin Roosevelt's "Fireside Chats" were vital in lifting the spirits of a Depression-riddled population and brought hope and confidence during World War II.

As radio matured, it became a medium of dreams. Young boys would sit up at night, listening as their favorite athletes competed, and dream of becoming the next Mickey Mantle; budding musicians listened and dreamed of recording stardom; and young actors and actresses dreamed of New York, and later Hollywood, as they listened to radio comedies and dramas. Broadcasters themselves became stars, and listeners came to think of them as friends they had never met. Everyone in a given community knew the names of their weatherman and newsman.

Over the years, radio has been both an expression and a reflection of humanity, where the good, the bad and the ugly all share the same stage. In every community, including Augusta, people still rely on radio much as they did before the advent of television or the Internet. Call letters sometimes come and go, formats change and broadcast booths become musical chairs as on-air talent comes and goes, but radio in the cyber age is every bit as relevant today as it was in the time of Charlie Chaplin.

In each of its seven decades, WGAC radio has been both an expression and a reflection of the Augusta community. The station's history makes WGAC a living time capsule of both the good and the bad that have happened to radio as an industry and as a social structure. The old WGAC studio building on Augusta's Broad Street is today an empty shell that casts a cooling shadow on the block of sidewalk where an impoverished young African American in the segregated South once played harmonica and shined shoes. The marquee at the corner of Seventh and Broad is long gone, but James Brown's footprints remain in front of what was once a marble radio showplace. The modern-day WGAC studios look remarkably like the studios that station founder J.B. Fuqua designed in the 1940s. Like the Broad Street building, Fuqua is long gone and his conglomerate, once known as Fuqua Industries, is a memory, but that first accomplishment he made as a young businessman (WGAC) remains, and it is stronger today than it has ever been.

For more than seventy years, the walls of the WGAC studios have absorbed the sounds of voices being broadcast to Augusta and the surrounding area. Those same walls have served as the incubator for men and women, both white and black, who would make great achievements in life. Harley Drew, a fifty-year veteran of the radio business, once commented that if "these walls could only talk," the stories they could tell would sound like fantasies. With this book, the walls of WGAC have talked, so to speak, for the Augusta State University students who have written these chapters have tracked down the stories of WGAC's incredible on-air personalities, business managers and staff. For seventy years, WGAC has made public service to the Augusta community its chief aim. You will read how that all came about.

This book is more than just a history of a scion of Augusta media. It is a collaborative project between the radio station and Augusta State University, and that gives it a historic flair as well. The idea for the book originated in 2010 when WGAC started running ads that celebrated its seventieth year on the air. ASU professor Debbie van Tuyll listened to those ads, and an idea for a project for her journalism history class came to her: the station needed a history written. And she knew just the person to approach: Scott Hudson, her former student who works as a reporter for WGAC. A collaboration was born, and the following fall, the twenty-two students in Debbie's journalism history class began working on this history. This book is the result of their labors, and a pretty decent result it is!

ACKNOWLEDGEMENTS

A ny project like this requires many thanks to be extended to those who have helped along the way. The greatest thanks must go to the men and women of WGAC, both its present staff and the giants on whose shoulders the station has been built. First and foremost, we must thank the station founder, the late J.B. Fuqua, and his wife, Dottie, for their generous gift of an evening talking with Scott about the history of WGAC. Much of the background he provided for this book was the product of that interview.

The present station managers and employees have also been amazing to work with. They have given generously of their time, fielded questions and concerns and trusted us to do our work and to do it well, even knowing that it will reflect on the station. Kent Dunn, the general manager for the Beasley Broadcast Group in Augusta, worked tirelessly, answering questions, loaning pictures and hosting station visits. Thanks also go to Austin Rhodes, Ashley Brown, Mary Liz Nolan, Harley Drew and Keith Beckum, who gave generously of their time. We would also like to offer special thanks to George Beasley, owner of Beasley Broadcasting, and his sons, Brian and Bruce, each of whom provided much-needed assistance and support for this project.

We must also thank Dr. Robert R. Parham, former dean of the Pamplin College of Arts and Sciences at Augusta State University, and Dr. Pamela Hayward, chair of the Department of Communications and Professional Writing, for their support of this project. Writing a book with undergraduates in the course of a semester is a daunting prospect, but both had confidence that the project would come to fruition and that it would be a quality product.

The WGAC family members, listeners and Augusta community members who shared their memories with us are ultimately responsible for this book coming to fruition. Without their help, this project would not have been possible. Thanks also go to the Georgia Radio Hall of Fame for allowing us to use photographs of George Fisher; to B.J. Wood for the use of his editorial cartoon; to Saturated Simplicity, Beasley Broadcasting, John Diskes and the Fuqua School of Business at Duke University for supplying photography; and to attorney Christopher Hudson for his legal advice, explanations of how conglomerate corporations work and willingness to help with editing chores.

Additionally, we would like to acknowledge the following students in Debbie's book production and editing class in the spring 2012 semester. They have each worked as diligent production editors on this project, and they have each taken the project seriously and given it their absolute best efforts. They are to be commended for their work.

Mary Anderson	Jacqueline Brill
Fox Si-Long Chen	Catherine Collingsworth
Alexandra Eash	Amber Forbes
John-Michael Garner	Jessica Hanson
Kelli Harris	Kristin Hawkins
Jasmine Housey	Amy Hunter
Patricia Johnson	Tamika Lampkin
Daphne Maysonet	Madia Orlando
Megan Petrie	Stuart Prather
Megan Waite Steinberg	Sara Tafazoli
Stenson Willard	Valerie Wooten

Chapter 1

FUQUA'S VISION

Chelsea Mathews, Lauren Kellums and Scott Hudson

All I had was a vision and an abundance of self-confidence.
—J.B. Fuqua

On June 26, 1918, John Brooks Elam Jr. was born into a life of tragedy and hardship. After losing his mother due to complications of childbirth at the age of two months, he was raised by his maternal grandparents on a tobacco farm in Virginia. He eventually changed his last name to his mother's maiden name, Fuqua, dropping his surname completely. Fuqua's childhood was consumed by working strenuous, long hours on the farm. Despite his arduous way of life, Fuqua was extremely persistent in taking the time to educate himself. This desire for knowledge was partly due to lacking children his age with whom to play. Being a lonely child was actually a blessing in disguise for the young boy; he used his ample free time and curiosity on the tobacco farm to read every book he could find.[5]

After depleting the small library of his local school, a favorite teacher advised that he could borrow books from the library at Duke University. After each long day's work on the family farm, Fuqua would read by candlelight and learn about things that interested him. He was intent on educating himself so that he would not have to live the hard life of a farmer, and he used what he learned about finance and business to devise ways to build an empire using other people's money. Later in his life, once he had accomplished that objective, Fuqua would repay Duke University's generosity with its library holdings by donating $10 million to endow the college's Fuqua School of Business.[6]

Fuqua's life was changed forever, thanks to the simple turn of a radio dial and twenty-five cents, when he was in his mid-teens. While sampling local radio stations on his family's first radio, Fuqua heard an announcement from the chief engineer of a Richmond, Virginia radio station. For the price of a quarter, listeners could receive by mail a book, *How to Become an Amateur Radio Operator*, that would teach them about the new technology. Of Fuqua's many business dealings, he insisted that this was the "greatest investment of [his] life." Over the next few years, Fuqua relentlessly gained an expansive knowledge on the art and science of radio engineering and technology and built his own ham radio set while teaching himself Morse code.[7]

After graduating high school, instead of pursuing a college degree, Fuqua attained his commercial radio operator's license. Shortly afterward, his dream of getting off the family farm became a reality. He obtained a job on a freighter as a radio operator and embarked on a journey that took him around the world. Unlike his shipmates, who spent the majority of their free time fraternizing in bars, Fuqua spent free time just as he had back on the farm: reading educational books to quench his undying thirst for knowledge, especially about finance and business. After a year on the ship, he was ready to move on to other radio ventures. Fuqua understood that radio could be much more than a navigational tool for shipping. He believed it could also be a serious moneymaker for an ambitious individual who was not afraid of taking risks. Fuqua knew that he was just the man to bring radio to the masses.[8]

When Fuqua got home from his maritime adventures, he used the money he had earned to pay for some formal education in radio. He enrolled in Capitol Radio Engineering Institute in Washington, D.C. After receiving his degree in radio engineering, Fuqua took a temporary job as a summer relief operator at WIS Radio in Columbia, South Carolina. True to his ambitious nature, after just a year, Fuqua started his move up the ranks when he moved on to a chief engineer position at WCSC, WIS's sister radio station in Charleston. Fuqua excelled in his

WGAC ran a full-page ad in the *Augusta Chronicle* on December 1, 1940, to announce that it was signing on as Augusta's second radio station and would be providing top-notch programming that included entertainment, news and the weather.

engineer duties, but he began to look at two things: how to make radio clearer and more pleasant to the ear and how to make some money off starting his own station.[9]

Before long, Fuqua realized that he had learned enough about the radio industry to make him more than capable of operating his own station. After researching different small towns, he decided that Augusta, Georgia, would be the perfect location to start a new station. Augusta had a central location, and its radio market was wide open. Only one low-power radio station, WRDW, served the greater Augusta area, so he made plans to visit to check things out firsthand.[10]

Fuqua arrived in Augusta not knowing a soul and barely having a dime to his name. But none of that discouraged the young radio engineer from going after his dream. Wearing the only suit he owned, Fuqua marched into the chamber of commerce and persuaded the secretary, Lester Moody, to give him the contact information for three of the wealthiest men in the area. Fuqua found the three men willing to invest in a new station by the end of that very same day. The station he founded with those men's money would become one of the most influential and recognized radio stations in Georgia's history. Fuqua would serve as general manager of the fledgling station.

Initially, WGAC was owned by Twin States Broadcasting Company. Local attorney Frederick Kennedy was president, *Augusta Herald* editor Milwee Owens was vice-president and *Herald* publisher Glenn R. Boswell was secretary and treasurer of the company. The involvement of newspaper principals in a radio station was unusual for the time, for newspapers were still wary of their new electronic competitors. In some communities, newspapers even blocked radio stations' use of Associated Press news copy or refused to run radio listings for fear of losing readers to the competition. That sort of rivalry did not exist in Augusta, however. The city's dominant daily, the *Augusta Chronicle*, carried radio listings for both WRDW and WGAC, and *Augusta Herald* newsman Sam Moss served as the station's news commentator. Each week, Moss would put together a fifteen-minute commentary on some news topic of interest to the station's audience.[11]

J.B. Fuqua always admitted to having a large ego, even saying in his memoir that he did not trust anyone without a healthy ego, but his wife, Dottie, laughed herself silly recalling how J.B. rented a car, drove to Augusta, demanded meetings with people he had never met and walked away owning part of a radio station. Fuqua, however, had done his homework, and he knew that starting up a new radio station in Augusta was more than a calculated risk. It was a potential gold mine. The Augusta market was aching for Fuqua's vision in 1940. With only one other local station in the small market, and only one strong statewide station broadcasting to the area (WSB in Atlanta), Fuqua saw an opportunity in Augusta that would eventually lead him to a net worth of several million dollars. The devilishly clever part of the equation was that Fuqua never put up a dime of his own money to start a station that became the foundation of a financial empire. WGAC started up with $10,000 in capital but not one penny from his own pocket.[12]

Just on the heels of the Great Depression, Augusta's economy was poised for strong growth in 1940. Business was picking up in the Garden City, as Georgia Howard Lumber Company brought new jobs and new opportunities to Augustans. World-renowned golfer Bobby Jones had gifted the city its now famous Augusta National Golf Course in 1933, and its first Augusta National Invitational Tour (now the Masters Tournament) happened one year later. Thanks to a seven-mile-long industrial canal off the Savannah River, Augusta—or "Little Lowell" as it was called, nicknamed after the Massachusetts mill city—provided numerous opportunities for businesses to succeed, thanks to the city's extensive manufacturing and power advantages.

Further, Augusta had two institutions that provided stable employment opportunities for local workers. The Augusta Arsenal, located on what is now the campus of Augusta State University, manufactured munitions that would be purchased by buyers preparing for a worldwide military conflict. Camp Gordon (today Fort Gordon) promised a full flow of military recruits to stimulate the local economy. Augusta also remained a hub in the southeastern cotton industry. The John P. King and Sibley Cotton Mills, located on the Savannah River in downtown Augusta, employed a combined 1,400 Augustans at the turn of the twentieth century. When other cities were still suffering from the Great Depression,

J.B. Fuqua saw the construction of the magnificent 1,500-seat Miller Theatre as a sign that the city was investing for its future.[13]

Fuqua, an entrepreneur with little formal education and an uncanny ability to persuade people to lend him money, also likely saw Augusta's medical district as a promising prospective location to find radio listeners who had expendable incomes, a key element to selling on-air advertisements. Augusta's Medical College of Georgia drew educated upper-class students to the region, and the established hospitals and other health practices helped to keep them there well into their adult lives. From any perspective, the stage was ripe for a man like J.B. Fuqua to enter Augusta and take a leading role.

Augusta's promising economic prospects, coupled with the city's central location near large cities such as Columbia, Atlanta, Alabama and parts of northern Florida, made the Garden City appear to be a prime spot for investment.[14]

WGAC debuted in 1940 at 1210 on the radio dial. It was the new kid on the block in a state in which commercial radio had existed since 1922. The state's first commercial station, WSB ("Welcome South, Brother"), was also the South's first station. Its frequency was strong enough that it could sometimes be heard up and down the East Coast and well into the Midwest. Augusta had gotten its first station, WRDW, not long after WSB went on the air. WRDW was started as a local alternative to WSB, but it was a low-power station whose signal could not reach far outside Augusta. WSB was the flagship station of Augusta, Georgia, and the entire South at the time, but WGAC challenged the Atlanta station for those honors when it moved down the dial to the AM-580 frequency almost four years to the day after it first signed on. Augusta listeners flocked to the new local station, and in no time, WGAC had more hometown listeners than WSB had been able to claim at the height of its popularity.[15]

WGAC signed on the air for the first time at 5:00 p.m. on December 1, 1940. That evening, the new station received accolades from *Behind the Mike*, an NBC program that originated in New York, and it also received a celebratory welcome from Augusta. At 6:45 p.m., Augusta's other radio station, WRDW, broadcast a salute to its new rival in a program that was simulcast on both stations. According to a report in the *Augusta Chronicle*,

"Representatives of WRDW will voice welcomes to the new station, and a special program of music will be given." The program was staged at the new Miller Theatre, and performers included WRDW's organist, Paul Reed, singer Ellen Ball and marimba player Charlotte Mills.[16]

A pantheon of local dignitaries was also on hand to welcome the new station. Augusta mayor James M. Woodall spoke, as did North Augusta mayor R.B. Mealing—the latter told listeners and those in the audience at the Miller that he was delighted to have a new radio station in the community and praised the advantages that a second station would bring with it. WRDW station manager W.R. Ringson also gave a warm welcoming speech.[17]

"In the year that the radio industry celebrates its twentieth birthday and the year that your station WRDW passes its tenth year of broadcasting service to its trade territory, it is my pleasure, speaking on behalf of the Augusta Broadcasting Company and its employs, to offer my felicitations to the newest addition to our great industry—radio station WGAC," Ringson said. Radio, he reminded the audience, was an extremely important medium because of its functions as a source of public information and service and a means of entertainment.[18]

The new station would run its own programming, as well as that of its network, NBC Blue, which would become ABC in 1943. In an antitrust action that year, the Supreme Court forced NBC to divest one of its second networks, NBC-Red. NBC-Red would become the present-day NBC, and NBC-Blue would eventually change its name to the American Broadcasting Company, or ABC.[19]

WGAC's ten-member staff included Fuqua as general manager, B.M. Dabney as advertising manager, John Watkins as program director and William N. Nungesser as chief engineer. The station also employed three announcers, Clarence Levy, John E. Lindsay and Bill Huck; two engineers, Gill Strauss and Murrell Prince; and a receptionist, Mrs. M.M. Spencer.[20]

The station took out a full-page ad on the day of its inaugural broadcast. The ad promised programming that included music, news, comedy and drama. The ad assured potential listeners that "[n]o money or effort has been spared in making Augusta's newest broadcasting station the finest

WGAC's second studio was located in downtown Augusta at the corner of Broad and Seventh Streets. *John Diskes.*

and most modern in the South." The station was equipped with the latest technology and the best network programming and news available.[21]

In December 1940, when WGAC was born, Fuqua was twenty-one. He was the youngest radio station manager in the country. He continued to expand his station by staying ahead of his competitors both technologically and socially. Because WGAC had stronger programming and the incredibly business-savvy Fuqua managing it, the station quickly gained the dominant position in the Augusta market. It also became lucrative after day one. The station rapidly built a reputation for being the most progressive station of the time.[22]

During WGAC's early years, the station tiptoed around corruption in local politics, such as the dealings of the Cracker Party, and other controversies that might cut into profits. Fuqua admitted to this oversight in his memoirs. He recalled that WGAC refrained from shaking too many trees or looking under too many rocks. The station did, however, become heavily involved in one illegal activity, though neither Fuqua nor the station ever profited from it.

The activity was known as "the bug." Working citizens would gamble their wages with coworkers on the stock market. Before radio, stock prices only ran in the morning newspaper, but when WGAC started carrying a midday stock report in January 1941, listeners could hear the stock reports at 12:20 p.m. and use them to make more informed decisions about which stocks to bet on so they could try to double their weekly salaries.[23]

Former newspaper boy Jim Brennen recalled that he would deliver newspapers to an area of town known as "Frog Hollow." Almost none of Jim's clients took the paper to read the news of the day; they took it to look at the stock numbers. In a sense, "the bug" was an illegal lottery of sorts. Those who bet correctly on the next day's stock numbers won the bug, which was only a few dollars. While Brennen never knew Fuqua and was just a newspaper carrier, he once mused that the newspaper industry suffered greatly since those stock numbers could be heard on the radio for free and so Augustans did not have to buy a paper to play the game.[24]

WGAC also benefited from being an NBC Blue affiliate. NBC Blue operated as a radio production and distribution service from 1927 to 1945. It was a sister content distributor to NBC's second network, the Red network. At the time, NBC Red was more familiar to listeners, and WGAC's affiliation with NBC Blue was a bit more exclusive. This gave WGAC the ability to air popular NBC Blue network soap operas, one of the most popular forms of radio programming among female listeners. The bulk of WGAC's advertising at the time was geared to the female buyers.[25]

One thing WGAC's managers had not excelled at was choosing a location for the station. The station headquarters were located in a cabbage patch just off Sand Bar Ferry Road in south Augusta. In those days before air conditioning, station employees were inundated with the putrid smell of rancid cabbage. Announcers literally got choked up on the air. Fuqua quickly came to two realizations. The first was that he was going to have to move the station. The second was that he would never be able to eat cabbage again. And he never did.

```
┤12:05—Sign off.

WGAC
P. M.
  5:00—WGAC Dedication.
  5:30—Behind the Mike (Salute to
     . WGAC), NBC.
  6:00—Vladimer Brenner, NBC.
  6:45—Salute from WRDW.
  7:00—Dinning Sisters, NBC.
  7:15—European Situation, NBC.
  7:30—Radio Guide Program. ●
  7:45—House of Dreams.
  8:00—This Curious World, NBC.
  8:30—Champions.
  8:45—Program Preview, NBC.
  9:45—Clyde McCoy.
 10:00—To Be Announced, NBC.
 11:00—Inauguration of Mexican
     . President, NBC.
 12:00—News.

   Use of a purebred beef-type bull
```

WGAC first appeared in the *Augusta Chronicle*'s radio listings on December 1, 1940.

Once he had made up his mind, Fuqua started looking for new studio space. He found it on Broad Street in downtown Augusta, not far from WRDW's studios. The new location of the radio station was far more satisfactory.

By all accounts, WGAC was a success. It was such a success that in 1944, Fuqua, by then a part owner of the station, petitioned the FCC to move the station to an even more powerful location on the AM dial: 580 kilocycles, with an operating power of five thousand watts. Fuqua had big plans for the station. He told the FCC that he intended to install the new five-thousand-watt transmitter in Martinez, a suburb of Augusta, just as soon as his application was approved and war exigencies allowed him to obtain the equipment. Fuqua predicted that the increased power supporting such a low frequency signal would give WGAC a larger coverage area than a fifty-thousand-watt station at the other end of the spectrum. As Fuqua explained to the *Chronicle*, "Contrary to the layman's ideas, place on the dial of frequency is much more important than power in obtaining coverage." Not only would the station's tower be the highest point within thirty miles of Augusta, it could also be used for television broadcasts, Fuqua told the paper. He predicted that the increased power and new signal would allow the station to reach 1.5 million people.[26]

His success at WGAC persuaded Fuqua to expand his radio interests, and in 1946, he established a new station in Charleston, South Carolina, WFAK, again using other people's money. That station is still broadcasting, though its call letters have changed to WLTQ. Three

WGAC
1240 K. C.
MONDAY

A. M.
5:55—News.
6:05—Dixie Dawn Club.
7:00—Morning News Roundup.
7:15—Checkerboard Time.
7:30—The Cheerup Man.
7:45—Drive-By Time.
8:00—News Here and Abroad, NBC.
8:15—Ross Sisters NBC.
8:30—News.
8:35—Hollywood Headliners.
8:45—Christmas Carols.
8:55—News.
9:00—The Breakfast Club, NBC.
10:00—Clark Dennis, Songs, NBC.
10:15—Helen Hiett—Today's News, NBC.
10:30—A House in the Country, NBC.
10:45—Prescott Presents, NBC.
11:00—News.
11:04—1240 Club.
11:55—Noon News.
P. M
12:00 Hill Country Ballads.
12:15 1240 Club.
12:30—Curbstone Comments.
12:45—Luncheon Dance Music.
1:15—Between the Bookends, NBC.
1:30—Hawaiian Harmonies.
1:40—Air Base Reporter.
1:45—News, NBC.
1:50—Local News.
1:55—Interlude.
2:00—Vincent Lopez orchestra, NBC.
2:30—Into the Night, NBC.
2:45—Care of Aggie Horn, NBC.
3:00—Daily Devotions.
3:15—Let's Pretend Ballroom.
3:55—News.
4:00—Club Matinee.
4:15—Rhythm at Random.
4:45—Club Matinee, NBC.
4:55—News, NBC.
5:00—Tenth Annual Diplomatic
      Children's Program, NBC.
5:45—Kaybee hit tune.
5:50—New River Boys.
6:00 Santa Claus.
6:30—Jam for Supper.
6:45—Lowell Thomas, NBC.
7:00—News.
7:05—Sports News.
7:15—Lum and Abner
7:30—NBC Concert orchestra, NBC.
7:45—To Be Announced, NBC.
8:00—Ray Herbeck orchestra.
8:15—"Melody Gems" — with Evelyn
      Nageley.
8:30—Chuck Wagon Boys.
9:00—National Radio Forum, NBC
9:30—For America We Sing, NBC.
10:00—Monday Merry-Go-Round—NBC.
10:30—Fight—Ray Robinson vs. Marty
      Servo, NBC.
11:00—Sammy Kaye Orchestra, NBC.
11:30—Glenn Miller orchestra, NBC.
12:00—News, NBC.
12:05—Sign off.

WGAC's programming on its first anniversary was much more extensive than it had been a year before. *From the* Augusta Chronicle, *December 1, 1941.*

25

years later, Fuqua sold his 10 percent interest in WGAC and bought 100 percent of Augusta's WTNT station for $75,000. He eventually changed the station's call letters to WJBF.[27] True to form, not only would Fuqua make WGAC a banner station, but WTNT/WJBF would be a trendsetting station as well.

In 1950, Fuqua followed up on another one of his profound premonitions on technological trends. He decided that television was going to become a fixture in almost every home in America. On Thanksgiving Day 1953, WJBF-TV went on the air. At first, programming was limited to only about twenty-five hours per week, but that was soon expanded to full time as more and more sponsors signed on.[28]

In 1960, renowned Augusta broadcaster Jim Davis was hired to do news for the station. Davis became close friends with Fuqua during his tenure at the station, but that did not mean that Fuqua let the newscaster slide on anything. Davis said that the best description of the J.B. Fuqua he worked with was, "Tough, but fair. Fuqua was a taskmaster, yet he respected and cared for all of his employees," Davis explained. He would even go out of his way to help them out when he could.[29]

When one of Fuqua's hardest-working secretaries was planning her wedding, Fuqua pulled a few strings and was able to host her ceremony on the grounds of the Augusta National golf course. The employee, Winona Warner, not only became a lifelong friend of Fuqua's, she also became the first female chief officer of a Fortune 500 company, Fuqua Industries. Warner broke through the glass ceiling at a time when women were expected to have dinner on the table when the family got home each evening and the bathroom cleaned properly. Warner was just one in a long string of high-performing employees Fuqua helped rise to the top regardless of gender or race. Fuqua also instructed Davis to hire Frank Thomas, the first African American television reporter to report for a white-owned station in Augusta.[30]

Fuqua had heart, but when Davis was working with him, he was not one to cut up. Business was a serious matter for Fuqua. As he aged, though, that would change. Perhaps finally achieving the success he had strived for all his life helped Fuqua see the lighter side of life, or perhaps the depression he had suffered from later in life lifted as he got older.[31]

Aside from being a taskmaster, Fuqua was also a continual pacesetter throughout his varied career. At WGAC, he developed some of the same sound-dampening techniques still in use today, and WJBF television was one of the first in the nation to switch to color programming. But Fuqua was not content just to build radio and television stations. He put together a business conglomerate that included oil-drilling endeavors, Snapper lawn mowers, Royal Crown Cola and a massive chain of movie theaters. Many today consider him the father of the conglomerate corporation. Fuqua was also smart enough to know when the conglomeration technique had reached its peak and change his strategy. Whatever business he was dealing with, Fuqua seemed to have an instinctive understanding of how to do whatever needed to be done to make it a success. His instincts helped achieve his childhood dreams of becoming a rich man. When he died in 1995, he was one of the richest men in the state of Georgia.

Fuqua was more than just a tycoon, though. He was a philanthropist, a teacher, a father, a husband and a politician. In 1975, a young politician by the name of Jimmy Carter stopped by Fuqua's office, seeking advice on his political aspirations. The former Georgia governor concisely stated to Fuqua that he was going to run for president. As the then chair of Georgia's Democratic Party, Fuqua advised the future president in the right direction. Carter indeed became president of the United States and attributed that victory to J.B. Fuqua on countless occasions.[32]

Having made many monumental and fortuitous business deals, Fuqua still contended that the best deal he ever made was marrying his wife, Dorothy "Dottie" Chapman, to whom he referred as his Madame Queen in the foreword of his biography.[33]

Fuqua commended his wife for being a great help in his achievements. They were married during World War II and had two sons, Rex and Alan. Tragically, Alan was killed in 1971, spurring Fuqua's battle with clinical depression. There were days when the mental despondency would immobilize Fuqua and hold him hostage in his own mind, in turn preventing him from making crucial business deals.[34] However, Fuqua rose from the ashes and overcame his debilitating disease.

Transcending his impoverished upbringing and surmounting the many trials and tribulations throughout his life, J.B. Fuqua was able to build a legacy. He received countless accolades and awards for his incredibly benevolent charity work and business innovations. His perseverance and clairvoyant business decisions made him a pioneer of his time. He took twenty-five cents and made it into one of the most successful business conglomerates in the entire world.[35]

Chapter 2

THE WAR YEARS FROM DOWNTOWN AUGUSTA

Ron Hickerson and Daniel Barber

As Fuqua and his new partners were working to get their radio station up and running, tensions were mounting across the globe. In May 1940, Germany invaded Belgium, France, Luxembourg and the Netherlands. Before WGAC could hit the air, Hitler had met with Mussolini to make war plans. German planes ran raids on Coventry, Southampton, Birmingham, Bristol and London, and the Japanese were already beginning to plan their assault on Pearl Harbor.

Most radio stations and newspapers in the United States covered the international tension, though in 1940 and 1941 most maintained a neutral stance toward the conflict raging in Europe. Americans had already been through one bloody war. No one was anxious to get pulled into another one.

That did not mean that Augustans ignored the plight of European peoples who were caught in the midst of warring armies. When German bombing raids on London rained destruction on the British capital in October 1940, Augustans created a British War Relief Society. Even though he was a newcomer to Augusta, Fuqua was invited to serve on the society's executive board, right alongside prominent Augustans like *Chronicle* owner W.S. Morris.[36]

In WGAC's first year on the air, the war would spread to Africa, Asia and the Middle East. Rommel would begin his operations in Tripoli. The Nazis would invade the Soviet Union, and they would also begin experimenting with their gas chambers at Auschwitz. America managed to stay out of the war that year, but it did agree to supply materiel to the Allies. A week after WGAC celebrated its first year on the air, however, Japanese bombers swooped in on a surprise attack, killing more than two thousand American servicemen and reducing Pearl Harbor and the U.S. Pacific fleet to rubble.[37]

The fledgling station had performed well in its first year. Its programming had attracted enough listeners that advertisers were willing to pay for sponsorships and spots, and it had been able to keep its bills and salaries paid. In fact, only a few months after Pear Harbor, *Life* magazine singled out WGAC as America's quintessential radio station. But hard times were coming for radio, the magazine predicted. Radio had been successful. It had been swell, *Life* agreed. "There is plenty that is fine and honest about radio," the article's unnamed writer contended. The new medium had provided "pleasure and education to the owners of 36,000,000 radio sets." Maybe radio had gone too far in playing to "the lowest common denominator," but it had made money doing so.[38]

That was all about to change, though, the magazine predicted. With U.S. involvement in the war, rationings and diversion of cash to war-related aims, businesses would no longer be able to sustain their high levels of radio advertising, the article continued. Further, the war had already brought a whole new set of problems to radio: how to sift fact from propaganda, how to keep crass commercialism from capitalizing on America's patriotic fervor and how to keep the public informed about events happening so far away and in such foreign places. So far, radio had not done so well, in the writer's estimation. Much of the medium's programming had consisted of "lamentable" jokes, "freak programs" and "washboard weepers," the magazine's phrase for soap operas whose heroine was "usually a noble woman who tries to straighten out other people's problems." The article dryly noted, "Nobody ever knew the trouble a soap-opera heroine sees. The episodes are drenched in tears, agonies, complicated misunderstandings." Music programming was the one thing radio had gotten right, according to the magazine's assessment. Radio used "its fine musical achievements to cover up a multitude of its other sins."[39]

Most of those sins were committed on small-town radio stations, according to the article. Of the nine hundred radio stations then operating in the United States, the majority broadcast with power of less than 1,000 watts. WGAC, like those other stations, was local and restricted in its programming, but that was why it mirrored the industry. "WGAC in Augusta, Ga., a 250-watter, is typical," the story explained. During daytime hours, its broadcast radius was about sixty miles. At night, when it had to cut power to make way for the nation's big clear-channel stations, its broadcast radius fell to only about thirteen miles. Its programming was also typical, the article continued. "Its local musicians, preachers, commentators and gossip columnists give a miniature image of the whole radio business." The station broadcast from 6:30 a.m. to midnight each day, and it filled those seventeen and a half hours of programming with content from the network as well as local programming.[40]

The article did not explain how WGAC came to the attention of the *Life* magazine staff, but its managing editor was John Shaw Billings. Known as the "Man Who Made *Life*," Billings was actually a native of the Augusta area. He had grown up eleven miles outside Augusta at his family's plantation, Redcliffe, located near Beech Island, South Carolina.[41] As the managing editor of a nationally circulating magazine, Billings and his wife were celebrities back home in Augusta. Their visits were always reported in the *Chronicle*.[42] Billings was also a member of the Richmond County Historical Society and was involved in matters in the Augusta area.[43] Billings and his wife came home frequently, and it could well be that they listened to WGAC when they were home and liked the station enough to feature it in the magazine.

Life magazine had a circulation of 3 million and claimed that more than 20 million people saw the pictures published in the magazine. It was "the largest weekly publication in the country" in the 1940s. Given the circulation of the magazine, the *Chronicle* estimated that "the space devoted to the Augusta station is equivalent to $25,000 worth of advertising space and is one of the biggest publicity 'breaks' the city has ever received."[44]

Daniel Field soldiers preparing for a boxing match. *Library of Congress.*

The magazine article itself was too short (only 123 words) to detail much about WGAC's programming, but *Life* rarely told its stories in words. It was the pictures that did the magazine's talking, and that was certainly true for the WGAC story.[45] The two-page spread, shot by *Life* photographer Alfred Eisenstaedt, detailed in pictures virtually every component of the station's typical broadcast day. The photo directly underneath the story showed Reverend Hamilton West of St. Paul's Episcopal Church preaching. St. Paul's services were broadcast live each Sunday morning on WGAC. Next to Reverend West was Betty Smith, a student at the Junior College of Augusta (today Augusta State University). She had a weekly program of campus gossip. The caption under the picture commented, "WGAC has daily programs of more serious local news, plus regular A.P. news casts."[46]

The pictorial spread also included a shot of a reporter in the field as he interviewed John Speer, a local stockman. According to the caption, cleverly headlined "Local Bull," WGAC devoted a good bit of news coverage to the area's livestock industry. Other photos included a sales meeting at which twenty-four-year-old station manager J.B. Fuqua was presiding, as well as a trio of high schoolers from Thomson, Georgia, who did some singing at the station: Barbara Burch, Lucy Lockett and

Winona Colton. That caption observed, "They sing in close harmony just like the big-time girl trios." The spread also featured a picture of three local brothers who provided a musical program for the station. They were flutist Antonio Pinaros, harpist Edward Pinaros and violinist James Pinaros. The headline for that photo was also a bit cutting: "Local brothers play 'good' music."[47]

Only one picture in the whole pictorial gave any hint that the United States was at war, and that was the second picture on the second page of the spread. At first glance, it appears to be just another picture of another big band so popular in that era. A closer look, though, shows that the band members are wearing military uniforms, and the music stands for the four alto saxophonists give the name of the group as "12th Inf." To the right of the band, four soldiers stand huddled around a microphone. Augusta in 1942 was home to Camp Gordon and a military airfield, Daniel Field, and WGAC had developed a weekly program for those soldiers that featured a military band and "an air field Winchell, bubbling over with camp gossip." That reference was no doubt an allusion to Walter Winchell, a popular national columnist of the day.[48]

The pictorial only hinted at what was a reality for American radio stations during World War II: listeners were completely caught up in the war. Virtually every aspect of their lives was influenced by America's involvement. The war loomed large for Americans, especially in places like Augusta, where Camp Gordon served as the headquarters for the Fourth and Twenty-sixth Infantries and the Tenth Armored Division. Later in the war, Camp Gordon would also become an internment camp for prisoners of war. Thousands of soldiers passed through there and Daniel Field, a military airbase where Jimmy Doolittle's pilots were trained. Doolittle's group would be responsible for the bombing raid on Tokyo.[49]

WGAC broadcast both network and local programming both for the soldiers who were stationed at the city's military facilities and for its Augusta listeners. NBC Blue offered more frequent news updates, and even entertainment content was flavored by the war. Each afternoon at 5:15 p.m., for example, WGAC would broadcast NBC Blue's fifteen-minute program *The Sea Hound*, which chronicled the adventures of Captain Silver and his sidekick as they searched for Nazis throughout the

Western Hemisphere. *The Sea Hound* was followed by another program, *Flying Patrol*, which was likely another war-related drama, though it has been difficult to find out anything more about it.[50]

The station also helped build support for the war through its programming. Its news programming was especially important because it was one of Augusta's only two broadcast outlets. Augustans could get news from only three local sources: the *Chronicle*, WRDW and WGAC. Once the United States entered the war, on any given day WGAC would broadcast 130 minutes of news. Of that total, 45 minutes was specifically war-related news. NBC Blue's London war correspondent, Morgan Beatty, was featured twice a day in two 15-minute programs each. One ran at 8:00 a.m. and the other at 10:30 p.m.[51] A year earlier, in the summer of 1941, a typical day would offer 75 minutes of news.[52] News programming nearly doubled once the war started.

News programming at WGAC meant both local and network news. Network news helped connect Augustans to national events, such as the president's Fireside Chats. These radio conversations gave the president the opportunity to pull the nation together as he discussed important issues, such as aid for Britain, the topic of his May 27, 1941 chat that ran on WGAC.[53] The speech was broadcast that evening at 9:30 p.m. and lasted until 10:00 p.m. It was carried on both WGAC and WRDW.[54] WGAC also frequently aired speeches made by other important figures at the time, including British prime minister Winston Churchill.[55]

While WGAC covered these important national events, its local news coverage was often lacking. At the time, "[t]here wasn't much local news coverage," wrote Fuqua in his memoir. "We didn't have teams of reporters around like radio stations do today." In fact, Fuqua said that the station's local news reports at the time consisted of headlines copied out of the local newspaper and read on the air.[56] WGAC's niche for local news coverage would not blossom until later, but the station was still maturing while growing in popularity. And WGAC proved its influential power in the Augusta area during the war through rallying volunteers to aid in the war effort.

WGAC encouraged local volunteering during the war in its broadcasts of *Red Cross Roll Calls*. The November 11, 1941 program, according to the *Chronicle*, was designed to "enroll every man and woman in Richmond County in the American Red Cross" in order to aid the American armed forces. [57] During the program, Augusta mayor James Wooddall encouraged citizens of Richmond County to enroll in the organization, reminding them that the American Red Cross was the only group capable of assisting and advising men in the armed services abroad and their families at home. [58] WGAC also encouraged volunteers by airing a speech made by Durice Dickerson, executive secretary of the Georgia State Nurses Association. The *Chronicle* reported that Miss Dickerson was invited to give an on-air speech about "the Nursing Council and its relation to Red Cross Nursing Services." [59] Her address also encouraged volunteering in the American Red Cross, specifically with regard to serving as nurses. The only Augusta station to carry the speech was WGAC.

The station did not just encourage citizens to volunteer; its employees stepped up to do their parts, too. Station Manager J.B. Fuqua was active in supporting war relief efforts, and he also worked on campaigns to support the sales of war bonds. Most of his war bond work involved running publicity campaigns to encourage people to buy bonds. One of his publicity efforts brought to town an exhibit of fifty photos of war fronts. The pictures showed Augustans real-life images of the war. According to the *Augusta Chronicle*, "The pictures are gripping in their portrayal of real battle conditions." [60]

The multiplicity of roles that arose for radio during the war gave it new power as a medium of mass communication, and that was a power the federal government wished to harness in support of the war effort. In fact, for a while, the federal government considered capturing radio's influence by establishing a government-operated network designed for propaganda purposes, but it concluded that the "four networks are now doing a better job than could possibly be done by one network." [61] Instead, government established several different offices to distribute government information. These offices were not objective news organizations. Instead, their job was to distribute information that would boost support for the war and help in recruiting men and women for necessary wartime work.

The first government information service was the Office of Facts and Figures (OFF). It was established after a private organization, the

Committee for National Morale—made up of psychologists, social scientists and media professionals—called for an American organization that would distribute information intended to build support for the war but without using the propaganda tactics of the Axis countries. The immediate response was the Office of Facts and Figures, headed by poet Archibald MacLeish. His mission was to disseminate information about the war widely so that citizens would understand what was happening. McLeish, a journalist and lawyer, worked from what he referred to as a "strategy of truth." He believed that citizens should have the proper facts before them rather than the lies and deception that were the hallmarks of German propaganda. That information, however, was presented from the perspective of the government. Some have also referred derisively to this sort of government information as propaganda, but a pair of Spanish scholars who studied the work of MacLeish and his organization concluded that his work—and that of the Office of War Information that supplanted OFF in June 1942—was an example of "the ethical contribution of propaganda to facilitate the dialog and debate which are necessary in democratic societies."[62]

The work of the various U.S. information services became critical after the attacks on Pearl Harbor in 1941. Their main assignment was to distribute war news for domestic use and promote patriotism through posters and radio broadcasts. The Office of War Information, a much more extensive organization than OFF had been, was responsible for formulating and executing "information programs designed to facilitate the development of an informed and intelligent understanding, at home and abroad, of the status and progress of the war effort and of the war politics, activities, and aims of the Government." With the responsibility of informing the American public, these government offices became critical players in the maturation of radio programming during this time, especially with the broadcast of radio dramas such as *We Hold These Truths* and *This Is War!*[63]

This Is War! was a unique approach to radio programming. The program debuted on Valentine's Day in 1942. The half-hour broadcast ran at 7:00 p.m. on Saturdays on all four radio networks (NBC Red, NBC Blue, Mutual and Columbia). The first series ran for thirteen weeks

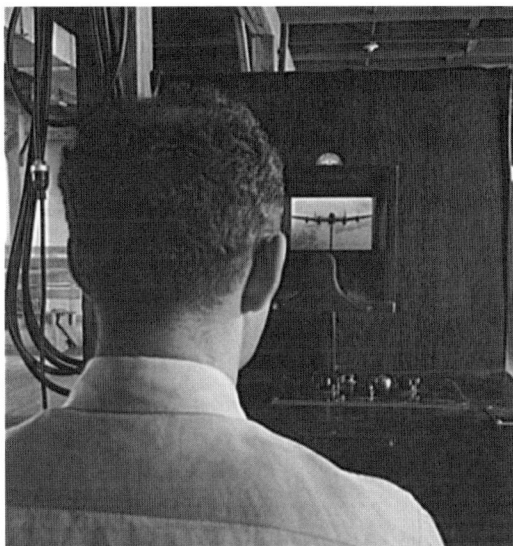

This page: Soldiers training at Daniel Field were instructed in aircraft identification. *Library of Congress.*

and was broadcast in seven foreign languages. The program was serious and hard-hitting. The host of the first episode, Robert Montgomery, a film star of the day, told listeners that the program was intended to be serious. "Laughter can wait," he said. "There's a war on." Directed by screenwriter and essayist Norman Corwin, the program was sponsored by all four American radio networks.[64] It ran on more than 700 of the nation's 924 radio stations.[65]

This program was influential in changing the way government information services reported war information. OFF, originally known for its dry reporting of statistics, soon shifted to reporting moving stories in the more entertaining fashion exemplified by *This Is War!* That program interlaced statistics, facts and news about the war with fictionalized characters and events. This helped dramatize the stories and allowed radio to capitalize on its entertainment qualities.[66]

Propagandistic in nature, the radio drama highlighted American values such as liberty, freedom of religion and private property and emphasized those values as reasons for America's involvement in the war. It depicted members of the Axis powers as the antithesis to the American way of life and argued that American soldiers were serving to protect fundamental human rights.[67]

The program took on different topics in each episode. Episodes dealt with the American soldier, the country's previous wars and women who stayed home. It encouraged mothers to rear good American children and continue to buy war bonds to support the war effort. The show pulled a good-sized audience, ranging between 19 and 24 percent of listeners in the first seven episodes, but the series was not an unmitigated success. Some Americans were uneasy with the propagandistic traits of the program, and President Franklin Roosevelt's opponents used it as the basis of political attacks.[68]

Special programs like *This Is War!* were not the only pieces of programming to take a patriotic bent during the war. Even the regular news at some stations and newspapers would romanticize the American soldier and other Allied countries while demonizing Axis forces.[69] The press used headlines such as "Reds Smash Stalingrad Attacks"[70] or "Soviet Juggernaut Pounding Nearer to Rostov: Reds Overrun Bitter Nazi Resistance"[71] to gain sympathy and support for the United States and its foreign allies. The news would also juxtapose the moral character and responsibility of the Allied soldier fighting for freedom with the depravity and lack of self-control of the Axis soldiers, who were depicted as heartless murderers who thought nothing of killing all the men in a village.[72] Radio news developed into a serious news medium during World War II.[73] Its purpose shifted from commentary and public affairs reporting to hard news, including the first eyewitness news roundups.[74]

These were national trends that trickled down to local communities like Augusta. Like its counterparts across the nation, WGAC also had the burden of providing information as well as entertainment to its listeners. It did so by airing network programs like *This Is War!* and *We Hold These Truths*, but it also did so with local programming intended to serve the needs of all those soldiers who were serving in Augusta, as well as the local audience. Throughout the war, WGAC would broadcast activities of the United Service Organization (USO) in Augusta. On October 13, 1942, for example, the station aired the Telfair USO program. The thirty-minute broadcast, aired from 9:00 p.m. to 9:30 p.m., included the top-notch performances of Eddie Lane and Jerry White, two Broadway performers who had enlisted in the army to help aid the war effort. The broadcast also included comedy acts from the Broadway duo and musical performances from the Fourth and Twelfth Infantries. This program offered variety shows to entertain citizens of the Augusta area and boost the morale of soldiers training for war at Camp Gordon.[75]

During this period, WGAC quickly achieved new popularity among Augustans by providing news and local and national entertainment. In its early stages of broadcasting, the new radio station grew rapidly. Part of WGAC's notoriety may have been due to the national notice it received from the *Life* magazine article in 1942, but there were other reasons, too. Its wartime programming certainly contributed to the station's growing reputation. Its affiliation with NBC Blue meant that it was able to air the latest news in a timely fashion for its listeners. For example, on June 6, 1944, NBC Blue London bureau chief George Hicks covered the events of D-Day via a recording that was provided to all of the American networks under a pool agreement. Hicks sailed with the crew of the U.S. Navy's *Ancon*. Early in the broadcast, atmospheric issues caused the recording to be marred with static, but Hicks managed to create a ten-minute report that described the bombardment taking place in detail.[76]

Another milestone occurred when Victory in Japan Day (V-J Day) marked the end of World War II after United States' bombings on

Nagasaki and Hiroshima. While there were premature broadcasts speculating on Japan's surrender the day before, the official announcement came on August 14, 1945, at 1:50 a.m. Eastern War Time. (In 1942, Congress enacted the War Time Act, which reinstated Daylight Savings Time as year round. The official designation "War Time" was used to indicate the year-round DST.[77]) The news was spread across CBS and NBC, along with reactions to Japan's unconditional surrender, which WGAC also broadcast.[78]

News continued to be an important component of WGAC's programming, even after the war ended. Racial tensions were heightened when veterans returned home to Georgia to job shortages and economic inflation. Blacks, who had by then helped the country win not one but two world wars, had grown weary of fighting for civil rights for others that were denied to them. Most black soldiers had expected to be treated with more respect after their service. They expected to receive the same postwar benefits as their white comrades. But prejudice and conservative white politicians, exemplified by Georgia governor Eugene Talmadge and Augusta's Cracker Party, meant returning soldiers had another war to fight.[79]

The Cracker Party was a long-standing corrupt and racist political machine that evolved out of the Democratic Party. To build support, its adherents did small favors for Augustans, such as cancellation of taxes, reduction of police-court sentences and adjustment of water bills. But there was also a darker side to the Crackers. *Augusta Herald* editor Milwee Owens, one of the co-owners of WGAC, claimed that the machine took 1 percent from each city employee's paycheck as a sort of kickback.[80]

But the Crackers were a taboo topic for most Augusta media outlets until 1942, when Bridge Evans, a local Augusta resident, was arrested for openly criticizing the Crackers at a high school football game. After the war, Augustans who opposed the Cracker policies formed a new Independent Party and put up candidates for office to oppose the party's entrenched politicians. The Cracker Party's back was broken in Augusta in the 1946 election. *Chronicle* publisher William S. Morris opposed Cracker Party boss Roy V. Harris for his seat in the legislature.[81] Morris beat Harris with 12,574 votes to 7,719. About 4,600 blacks voted in that election.[82]

Many of the Independent Party's white voters were former soldiers who sided with their black comrades and their struggle for equal rights. These white soldiers went against the entrenched political machine for two reasons primarily. First, they feared that the discrimination that kept black soldiers from receiving military benefits would lead to discrimination against southern soldiers in general. Second, after having served with and getting to know black soldiers on the front line, white soldiers began to see their comrades in more human terms. This mentality would permeate the minds of some Augustans—and thus WGAC listeners as well.

The Independents and Crackers would continue to contest local elections for several more years in Augusta. Finally, in 1953, the two parties worked out a compromise that allowed blacks to hold one or two wards in the city and thus obtain minority representation on the city council. A voting rights lawsuit in 1986 changed the way elections were conducted in Augusta and made it easier for blacks to be elected to city office. However, race and political power are unsettled issues in Augusta even today.

There is no doubt that World War II had a profound influence on the radio industry. On the national level, the war shaped radio into a formidable source of news with its rapid reporting and ability to deliver information almost immediately (by the standards of that time, live radio *was* immediate). Radio was no longer solely an entertainment medium. It had matured into a sophisticated information source where news took primacy. Some of the country's best journalists had been lured away from their print jobs into the radio business during the war. That gave radio news the kind of clout it had not had before the war.

The war also touched the city of Augusta, Georgia, and its new radio station, WGAC. The station was born into a period of hopeful optimism tainted by mounting tensions over events in Europe and Asia. WGAC came of age during this period and would follow the war through affiliate broadcasts from NBC and CBS. The station also felt the repercussions of the war as it met the huge responsibilities of radio. During this period in its history, WGAC endured its trial by fire and met it with success, establishing its identity as a popular radio station that served the needs of its listeners. In its programming, the station proved that it could entertain, inform and recruit members of the Augusta area. Its burgeoning popularity brought new influence, especially after the local station was singled out by *Life* magazine as "America's radio station."

When the war ended, former military men expected their financial and social situations to improve, and they fought to make those changes happen. While WGAC still had a lot more growing to do, the war years in Augusta acted as a springboard for the station's popularity and influence, establishing its identity as a grounded radio station and a formidable opponent to Augusta's other radio station, WRDW. But all was not golden for the station, for challenges lay ahead, especially as WGAC struggled to adapt to changing times that brought new problems for programming and competition.

Chapter 3

AUGUSTA'S TOWER OF STRENGTH

Jacquelyn Pabon, JoBen Rivera–Thompson
and Jillian Hobday

A stable economy in postwar Augusta, grounded in weapons production and cotton, provided a healthy advertising market for WGAC and allowed the station to focus on reaching a larger audience. With its strong new dial location at 580 kilocycles and operating power of five thousand watts, WGAC was poised to become not just a mirror of the industry but an industry leader, especially given the hard times radio faced in many places after World War II.

The war had been a boon to radio, despite *Life* magazine's gloomy predictions back in 1942. Radio had been the country's most important medium; not only did it provide a vital communication link between the war and listeners back home, it also brought live entertainment into households. Once the war ended, government programming dried up, and the importance of news receded; little that happened could match the drama and magnitude of war-related news.

Radio operators, performers and regulators were all casting about to figure out what the future of radio might be. Some thought that radio should be harnessed to serve social needs. For the most part, though, radio returned to its roots: entertainment.

In the midst of this postwar confusion, WGAC managers made the commitment to build state-of-the-art facilities that would take advantage of the new signal strength and put the station in the best competitive position possible. The newly obtained signal strength, coupled with the station's financial success, meant that it could start working on projects that would enable it to dominate its market.[83]

The first step in preparing to broadcast at the new frequency was the construction of a new tower and station building. Station managers began planning for a five-hundred-foot tower and its four antennas. The tower would be one of the highest man-made structures in the Southeast, and in mid-July 1946, the station encouraged Augustans to come out and watch the new towers being erected. "See how a slender steel tower is put up nearly 500 feet without scaffolds," an ad in the *Chronicle* encouraged readers.[84] The station's studios in downtown Augusta, at the corner of Seventh Street and Broad Street, were also being renovated.[85]

More than forty miles of wire had to be laid underground before WGAC's new tower was operational. The cost to construct the tower outside Martinez, a suburb of Augusta, was expected to run more than $80,000, but once that work was complete, the station was ready to sign on at its new frequency. WGAC was one of the first stations to be allotted a frequency so close to the bottom of the AM dial. This was extremely fortunate for the station. Its powerful frequency would become a key to the station's growth. The other key was the expansion of business in Augusta in the postwar period. Fuqua believed that a strong radio presence in Augusta would help the city grow, and he positioned his station to both contribute to and benefit from that growth.[86]

Planning for the new tower and studio had commenced in 1945, although the station would not reveal its plans to the public until February 1946. The plan, according to an ad in the *Chronicle*, was to make the station's downtown studios the "South's finest," with state-of-the-art design. "The NEW studios of WGAC will use the latest improvements in ACOUSTICAL DESIGN—POLYCYLINDRICAL WALLS and CEILINGS, never before installed South of New York or East of Dallas," boasted an ad. The studios would also feature the latest postwar broadcasting equipment. This would ensure unsurpassed fidelity of tone. "Our new studios are years ahead of anything

WHO'S L. M.?

THEY SAY HE'S A MEMBER OF "CLUB 58"

LEARN THE TRUE FACTS ABOUT THE MOST AMAZING EVENT IN AUGUSTA THIS YEAR — — — — — —

STUPENDOUS!   GIGANTIC!   COLOSSAL!

This enticing ad ran in the May 31, 1946 edition of the *Augusta Chronicle*. It announced the coming of a mysterious new "L.M.," none other than Lonnie Moore, who would become an important radio personality in Augusta.

LISTEN TO WGAC TOMORROW MORNING AT 13 MINUTES AND 13 SECONDS AFTER 11 O'CLOCK – – THE TRUTH WILL BE KNOWN – – –

L. M.? WGAC 11:13:13

in the South," the ad continued.[87] Fidelity in tone was a huge concern for the station. A quality tone meant growth in the quantity of the station's listeners. To improve that sound quality, the station invested in renovations to its downtown building that went beyond cosmetic to include acoustical alterations to prevent any echo at all. The change involved creating what were referred to as "polycylindrical walls" that were used throughout the building. WGAC would be the first radio station "south of New York or east of Dallas" to have this new technology, which used curved plywood facings for ceilings and side walls of the studio.[88]

Work was completed, and WGAC opened its $200,000 remodeling job to the public on Thanksgiving Day 1946. In a *Chronicle* article published on December 1, station managers emphasized their appreciation for listeners by inviting them to tour the new station.[89] About five thousand people poured into the station offices to see the "modernized" studio and its "splendid appearance." That same day, the station officially went on air at its new frequency broadcast from its newly redone studios.

WGAC's decision to revamp its studio and acquire a new, stronger frequency was not just a growth decision. Station managers were facing a threat from a new technology, one that was making people at the time wonder whether radio might be a sitting duck. That new technology had not only the power of sound but also the power of sight. It was television.[90]

A new frequency and physical location were not the only postwar changes that WGAC experienced. Programming was changing, too, thanks in part to the competition from television. Program length increased. The standard fifteen-minute time frame expanded to thirty minutes. Network programming continued to emphasize dramas, comedies and music. Quiz shows reasserted themselves, too. Quiz shows had been popular from the early days of radio, particularly on local stations. National networks were careful about them because of Federal Radio Commission (the precursor of the Federal Communications Commission) regulations related to lotteries and the licensing requirement that stations broadcast in the public interest. Quiz shows grew in popularity after the war, though. They offered a fun form of escapism, and the big money prizes were just too alluring for people who had endured years of austerity. By the late 1940s, though, many were beginning to question the ethics of quiz shows, and the FCC threatened nonrenewal of licenses for any station that broadcast quiz shows. This threat was eventually tested by judicial action, but by then the quiz show craze had run its course and radio listeners had moved on to other kinds of programs.[91]

Radio documentaries, earnest productions of an earnest time, gained new popularity in the postwar period. These special-edition programs typically focused on social issues such as alcohol abuse, juvenile delinquency and even effects of the atomic bomb, and they were big hits with postwar audiences. The popularity of documentaries gave rise to in-studio programs in which panels of experts would discuss the issues of the day. One of the earliest of these programs, *Meet the Press*, still runs today as a television program. *Meet the Press* premiered in 1945. Its producers were Martha Rountree and Larry Spivak.[92]

That same earnestness that gave rise to documentaries also influenced the nature of children's programming. New superhero characters appeared who fought injustice and wrongdoing. Most of them (but not all) were white. There were many Supermans and Joe Fridays, but there were also the Latino Cisco Kid and the Green Hornet's Japanese sidekick, Kato. Children's programs also commonly dealt with Cold War themes such as espionage and space exploration. Panic ensued when *American Mercury* magazine ran an article in its August 1953 issue claiming that Communists, who by then had been frozen out of the motion picture industry, were using radio to bring propaganda *"right in your own living room!"* (emphasis theirs).[93]

The new Red Scare ushered in by Wisconsin senator Joseph McCarthy in the early 1950s complicated radio's doldrums. Throughout the 1940s, the FCC had received complaints about the alleged pro-Russian views of certain radio news commentators. The commentators included noted newsmen such as Drew Pearson, William Shirer and Walter Winchell. Pearson was particularly targeted, for he and McCarthy were already bitter enemies and had gotten into a fistfight at a Washington, D.C., supper club in late 1950. McCarthy denounced Pearson as a Communist, and Pearson used his radio program to attack McCarthy. The FCC did little to interfere with broadcasting during the Red Scare. That may have been because internal investigations of FCC employees accused of being Communist sympathizers did not find support for the accusations, according to an oral history of the period done with then FCC member Clifford Durr.[94]

Much about radio had changed by 1950, but much had stayed the same as well. More than one hundred network radio series had been on the air for more than a decade. Most of the networks were virtually as old as the medium itself. Audiences were starting to get restless. They wanted new voices, programs and formats, and television offered a totally different approach to programming. Local stations had to find a way to respond if they were going to keep their audiences.[95]

WGAC had always prided itself on having the best on-air talent and programming in Augusta, but like national radio, something had slipped. Coming out of World War II and moving into the 1950s, programming on WGAC was staid, in large measure because it consisted primarily of programs from its national network. Largely, transmissions consisted of postwar-inspired news and popular dramas that got the Augusta listeners through their day. The station had state-of-the-art facilities, but it still needed to reinvent what listeners heard when they tuned in on the radio dial.

WGAC managers began to reflect on listeners' desires, and that seemed to pull them in the direction of music—the sexy, romantic and upbeat music of the big-band era, with suave crooners like Frank Sinatra and swinging melodies from band leaders like Count Basie.[96] Recorded music without interruptions, except for station identification, time, weather or news, became the standard nationally and at WGAC.

In 1941, WGAC typically ran just over three hours of music daily. By 1948, it was running nearly five.[97]

Much of that music was live and provided by the station's network. Sometimes, though, that live performance was in Augusta. On January 13, 1946, the station hosted Sammy Kaye's national broadcast. Kaye, one of the big-band era's most well-known band leaders, had brought his orchestra to Augusta to play for a dance the night before, but his daily national music broadcast still had to air. WGAC agreed to handle the broadcast from Augusta's Municipal Auditorium. Getting set up for the broadcast was an accomplishment in and of itself, according to an article in the *Chronicle*: "Before the radio program could proceed one step, this equipment had to be checked and set up, and telephone and radio lines tapped to feed the show into radio station WGAC." From WGAC, the signal went to Washington, D.C., and then on to New York. WGAC engineer Dave Freeman had charge of ensuring that the broadcast went smoothly. "Split-second timing is required," the article continued, "as the announcer and all commercials will come from New York. One minute's miscalculation would cause a jumble of words, mixed with music, ruining the program entirely."[98]

Despite the technical intricacies, only about forty-five minutes were required for WGAC's staff to move the necessary equipment into the auditorium. Rehearsing for the show, which was performed without a live audience, took longer. Kaye and his band "rehearse for hours before every broadcast," the article noted.[99]

As the Kaye event illustrated, music programming was becoming more and more important for radio. WGAC managers watched as stations across the country dumped serials and starting hiring announcers, later called disc jockeys, to spin the platters. Disc jockeys were personalities, entertainers who were more than just trained radio engineers who played records. Their appearance signaled the beginning of personality-driven radio.[100] WGAC's shift to more local programming led by popular local hosts again demonstrated its prowess for producing committed listeners by working to understand what audiences wanted from its radio station.

WGAC's final step in addressing the new realities of postwar radio broadcasting was to find those personalities that would help it keep its market share. Its first hire was Lonnie Moore, and the station heralded his new show with a newspaper ad that asked, "Who's L.M.?" The answer? "They say he's a member of Club 58." The ad did not explain what Club 58 was, but it encouraged listeners to tune in "tomorrow morning at 13 minutes and 13 seconds after 11 o'clock" to find out what "Stupendous! Gigantic! Colossal!" event awaited them.[101]

Moore became WGAC's best-known announcer of the postwar period and something of a celebrity around Augusta, especially with the older women. At least twice in the late 1940s, Moore hosted what were referred to as "Tom Breneman Parties." Tom Breneman was the popular host of a morning radio program called *Breakfast in Hollywood*. This was a loose-format show that was broadcast from Breneman's restaurant in Hollywood. The show featured audience interaction punctuated by music and film stars of the day. Notable features of a Tom Breneman Party included women wearing funny hats and the presentation of orchids to the oldest lady attending.[102]

Club 58, which catapulted Moore to that celebrity status, aired every weekday in the prime afternoon drive time slot, 3:00 p.m. to 5:00 p.m. It was a Top 40 hits show, and Moore would take requests and calls from listeners. He also sometimes did interviews with featured artists from a variety of music genres on his show. Sometimes those artists were not stars when they walked into Moore's studio, but they would be when they walked out. Don Rhodes, an entertainment writer for the *Augusta Herald* in 1990, recounted the story of the day when an unnamed gospel group from way down south in Georgia walked in to a live appearance on Club 58 and came out as the famous gospel/blues group the Swanee Quintet. Today, the Swanee Quintet remains the oldest all-black gospel group in the United States.[103]

The program was wildly popular with Augustans. WGAC received some of its highest praise and listenership for the variety of music and entertainment included in the Club 58 show. The content of the show was really not that different from what was running on stations across America. It featured the jazz and big-band swing tunes that were topping the charts. What was unique about WGAC's show was the host.

Personality-driven radio was the order of the day. If it was going to be successful, a show had to be hosted by an extremely personable, well-spoken and musically knowledgeable disc jockey. Moore was exactly that. Just a year before he came to WGAC, he had been a navy man serving on the Pacific front at the closing of the war. When he came to Augusta, he would pioneer the DJ role and perform it for more than a decade.

Moore was a stand-up guy with a clear voice, and his popularity grew to unprecedented heights for Augusta radio. He was considered Augusta's favorite radio personality. Beyond his on-air talent, Moore also became a presence in the community. He was not just a distinguishable voice with an unrecognizable face, like many jockeys; rather, he was actively involved in the community. He made appearances, hosted events and even charmed old ladies by giving them orchids.

As Moore's popularity peaked, radio faced another change: the rise of FM radio and its better sound quality. That change threatened WGAC's dominance in its market, but the station made the decision to stay committed to the AM band. With Lonnie Moore at the helm, the station was able to keep the hits spinning. Although Club 58 never left the AM band, the program remained popular enough that it still could attract visits from artists who would drop by to plug their latest recordings. One of those singers who may have made his radio debut during Moore's stint at the station was Augusta's most acclaimed musical figure, the Godfather of Soul, James Brown. Brown told an Augusta reporter that he had made his on-air singing debut at a Broad Street radio station and that he had a job outside that station as a shoeshine boy.[104] Brown also told that reporter that the station was WRDW, a station he would later own.[105] The problem with the story is that there was only one Broad Street radio station with a shoeshine business outside it when Brown as a boy, and that was WGAC.[106]

Moore was not the only star personality at WGAC in this period. Gruff-voiced, cigar-smoking George Weiss would found one of the country's top performing radio stations in 1947, but he got his start in radio working as a morning announcer at WGAC. He earned a whopping $22.50 per week. An unwavering entrepreneur, a celebrated news reporter and a ruler of the airwaves, Weiss was a legend in the radio world.[107] In 1947, he founded WBBQ, whose call letters represented the combined call letters of two stations from his hometown of Chicago, WMAQ and WBBM. That station would dominate the Augusta market

George Fisher, *right*, with one of his all-time favorite band leaders, Count Basie. *Cindy O'Brien.*

for thirty-five years, and it got its start in the same studios where WGAC had begun—the small building in the cabbage patch off Sand Bar Ferry Road—using equipment the former tenant had left behind.[108] WBBQ signed on the air on January 12, 1947, at 7:55 a.m. It was an affiliate of the Mutual Broadcasting Network. Listeners who tuned in to 1340 AM that day unwittingly witnessed the first broadcast of Augusta's most successful radio station ever.[109]

Almost immediately, WBBQ challenged WGAC for supremacy of Augusta's airwaves. Its sound was new, its programming—half news, half music—was unique. WBBQ's model called for news to be broadcast sixteen times each day to "keep Augustans up to date with the latest developments both at home, throughout the nation and abroad." Edward Dunbar, who was in charge of the station's news department, gathered some of news locally, but the station also carried news from Mutual, the Associated Press and the International News Service. WBBQ started its broadcast day at 6:00 a.m. and retired at thirty minutes past midnight.[110] Weiss and his station hit the open road to notoriety and left WGAC in the dust, at least for a while.

Despite their shared beginnings in the same studio space and the connection with Weiss, it would not be long before animosity moved in and built a wedge between WGAC and WBBQ. And just like in the movies, it all started with a conspiracy.

The story hit the newsstands on February 11, 1949. WGAC was suing WBBQ and Troy Agnew, owner of the minor-league baseball team in Augusta, the Tigers. WGAC had signed a contract with Agnew on February 2 to broadcast all of the team's out-of-town games. WGAC's suit alleged that on February 3, Agnew had sold the team to WBBQ in order to void the contract signed the day before. The suit alleged a conspiracy among Weiss and the other named defendants, which included W. Montgomery Harrison Sr., Dudley H. Bowen and Randall K. Strozier. Agnew and the bank manager who put up the financing were also named as defendants, but they were not accused of being part of the conspiracy. Fuqua's suit sought $10,000 from each of the conspirators and asked the Richmond County Superior Court to enforce his contract of February 2.[111]

Fuqua alleged that he had talked to Agnew about the broadcast rights and encouraged him to see what kind of deal WBBQ might offer. The next day, Agnew called Fuqua and told him that WBBQ's offer had not been satisfactory and that he was ready to deal with WGAC. He and Fuqua negotiated a contract, got it notarized and then went over to Fuqua's office so the station manager could write a $1,000 check as a "down payment" for the broadcast rights.[112] Purportedly, after that, Agnew gave WBBQ an attractive offer on the ball club, and the station bought it. WGAC's suit argued that the new team owner was bound by the terms of their contract with Agnew.[113]

In its defense, WBBQ argued that the suit was premature because baseball season would not start for nearly two months. Weiss's station also claimed that the Agnew-WGAC contract was for "personal services only, and could not be performed by WBBQ even if it was that station's desire."[114] The verdict: when purchasing the ball club, WBBQ did not assume the contract between Agnew and WGAC. The defendants paid $8,500 in damages. WGAC received the monetary award that was less than it desired ($40,000 punitive damages and $19,000 for a breach of contract), but it lost its bid to broadcast out-of-town games.[115] WGAC may have won the case on paper, but it seems that George Weiss's WBBQ won the main prize that sparked this brawl in the first place. Weiss's comment when the suit first was filed was, "Every knock is a boost."[116]

WBBQ eventually gained the no. 1 rating in Augusta, but equally important, it was highly ranked in the nation as well. Duncan's American Radio, a market research firm, reported that only five other radio stations in the country topped WBBQ in ratings for their weekly share in cities the size of Augusta.[117] Some have attributed WBBQ's success to its star announcer Harley Drew. Drew signed on with WBBQ in 1963 as a mobile news reporter and a fill-in disc jockey. "I worked there twenty-five years, during which the station never, ever lost a rating," Harley boasted. "Always no. 1."[118] Listeners agreed, or so the ratings seemed to show. From 1962 to 1990, WBBQ dominated the no. 1 spot in Augusta. Things started to slip for WBBQ, however, when a rating service, Birch/Scarborough Research Service, found it had dropped to the no. 2 in the Augusta market, coming in second to Top 40 station WFXA.[119]

During the 1990s, urban contemporary programming began to overshadow big stations like WBBQ across the nation.[120] The ascent of WFXA was attributed to changing tastes in Augusta listeners.[121] That's what WFXA and other Top 40 stations suggested to explain their success at bringing down Augusta's dominant station. According to those stations, change was what WBBQ was lacking, but Weiss was not convinced that anything had changed or needed to be changed. The rating service that showed the slippage was not Arbitron, the radio industry's standard ratings service. "Unless it shows up on the Arbitron, we still consider ourselves No. 1," Weiss told the *Augusta Chronicle*. He was not going change his programming to parallel the styles of Top 40 stations, at least not yet.[122] Before long, WBBQ was back on top. Ironically, Weiss thought it was the publicity the station received after "being no. 2" that brought in more listeners.[123]

In the mid-1990s, Congress adopted new ownership rules for broadcast media. Before, law forbade owning more than one radio station in a market. The change in rules in the 1990s opened up radio ownership by adopting a sliding scale that increased the number of stations an individual or corporation could own based on community size.[124] The result was considerable wheeling and dealing as broadcast companies bought and sold stations almost like they were trading cards. In 1992, WBBQ was the only radio station in Augusta that had not been sold, bought or traded.[125] Weiss attributed the "selling frenzy" in the radio market to the recession. Less money in the economy meant less advertising; less advertising meant less money to keep a radio station afloat.[126]

The tough times had another effect, however, and that was to convince Weiss that it was finally time to shift formats. In the 1990s, WBBQ changed its format to Top 40. Weiss had witnessed the success of Top 40 programming by observing the New York station WABC. Drew remembers the day Weiss set aside his pride and crossed over. Weiss called down to the station and finally said, "Go Top 40!"[127] He began sneaking in a popular song every half hour, and he bought two of the strongest radio signals in Augusta and "did what most people thought was ratings suicide." Weiss mixed Top 40 in with news radio, but that did not last long, for the station faltered in finding a new niche for its new market conditions—that and Weiss was sick.[128] In 1997, Weiss donated WBBQ and its sister station, WZNY, to the then Medical College of Georgia shortly before his death from cancer. MCG, today Georgia Health Sciences University, sold the stations to Cumulus Broadcasting for $14 million, which was used to establish the Weiss Endowment for Cancer Research. Today, news is only a small component of WBBQ's programming, and its format has changed to adult contemporary.[129]

Like its chief competitor WGAC, WBBQ had its share of peaks and valleys. One of the valleys came on May 6, 1973, between 4:00 a.m. and 8:00 a.m., when rain nearly submerged the station's transmitter. The station was located next to a storm basin, and the basin had filled up with trash and was not able to divert water properly. Weiss appeared less concerned about the transmitter than the "brand new beautiful 'gorgeous red shag carpet of the highest quality, just installed'" in his office that he found under four feet of water. Weiss and the WBBQ crew walked around in rubber galoshes, salvaging files and equipment, but they were not able to save the shag carpet. Weiss could only joke about the incident. "If we had known nine years ago, when we leased this property, that this situation would occur, the building would be on stilts," he told the paper.[130]

Weiss was one of Augusta's great newsmen and radio operators. He was known by many, but all knew him differently. To the medical community, he was a generous philanthropist who supported medical research. To other area stations, he was the owner of a dominant station that controlled the Augusta airwaves. He was the man who owned the station that they constantly competed against and strived to become.[131] To employees like Harley Drew, Weiss was a stickler for accuracy. At times, he was a "grouch" who would "come up behind you in his big news vehicle with a loud speaker and say, 'I can tell you're not wearing your seatbelt!'" Drew recalled. Employees rarely received compliments on their work, and if they did, it was merely a slip-up.[132] To journalists who interviewed him over the years, he was a witty and humorous man with stories about jacking up transmitters and wearing rubber galoshes. To his listeners, he was the voice behind the news.

Though Weiss was the consummate businessman who was able to build his radio station into a competitive giant, his first love was not business. It was news. Multiple times a day, his recognizably rough voice made its way across the airwaves and into the cars and homes of many—actually, most—Augustans. No matter the news, no matter where, no matter the time, Weiss was always there describing the scene by a two-way radio.[133] From the early 1950s to the late 1990s, WBBQ dominated radio news. George Weiss was "Car One" of WBBQ's white fleet of news cars, which were adorned with flashing orange lights. Mobile news quickly built up ratings and increased listener loyalty, making WBBQ a leader in news reporting and "an icon in the industry—a ratings powerhouse and market leader."[134]

Along with the loss of George Weiss was the death of mobile news. New owners of WBBQ, Cumulus, ditched the legendary white cars in 1998. News did not contribute to the bottom line, and so the new owner emphasized other types of programming. Listeners could no longer tune in to WBBQ to find out if the traffic they were caught in was due to an accident or just the product of the usual rush-hour delays.[135] WGAC's people believed that Cumulus dropped mobile news because it was new to the community; the company had little idea what this meant to Augusta.[136] This shocking loss marked "Augusta's sobering introduction to corporate radio."[137]

But Cumulus's decision also opened a window for WGAC to reassert itself in the Augusta market. In short order, WGAC adopted the same mobile news approach that had been WBBQ's hallmark and started providing on-the-scene news reporting for the city. Kent Dunn, who was the general manager of the Beasley Broadcasting Group Inc. in Augusta, made the decision to step up WGAC's news product because he knew that mobile news was a unique brand to Augusta. He saw the potential of new listeners and better ratings with an already established news product. "When they dropped it, it was a perfect opportunity for us to steal the news-talk brand for GAC," Kent said. "It helped us solidify what the format is and what it means to Augusta [today]."[138]

Those white cars do not rack up the miles like they used to during Weiss's days, but they are still on the road in Augusta. Obtaining the mobile news product was a quick jumpstart and a smart move by WGAC, for it put the station in the position to build on some other opportunities that had cropped up in the 1990s, opportunities brought about by the rise in the popularity of talk radio.

Chapter 4

THE RISE OF ROCK AND THE DECLINE OF WGAC

John–Michael Garner, Travis Highfield, Stephanie Hill, Karl Frazier, Rashad O'Conner and Armani Grant

All the old black-and-white reruns of 1950s television shows depict that decade as an idyllic time in American history when dads drove off to work every morning in the family's only automobile, moms cooked dinner in pumps and pearls and children were just the tiniest bit naughty, which only made them that much more endearing. Viewers could always count on the good guys—Superman, the Lone Ranger or Sergeant Joe Friday—to win the day. As the decade gave way to the 1960s, though, darker clouds gathered on America's horizon. The Cold War heated up with the Soviet Union's launch of *Sputnik*, America became embroiled in new wars "over there" to prevent the spread of communism and a junior senator from Wisconsin personally ushered in a new Red Scare that turned American against American. At the same time, manufacturing and technological advances made more goods available, and for the first time in decades, Americans had the money to afford those little luxuries like an electric iron or record player. The period's uncertainties and anxieties, coupled with growing affluence, brought about a cultural change that affected almost every component of American life, including music radio.

Those same young people who had been sock-hoppers in the 1950s were at the vanguard of this cultural change. They rebelled against what many considered the repressive strictures of the 1950s and sought a new form of collective identity, one that focused on music. Further, because of the postwar baby boom, the sheer number of young Americans was vast. Business recognized the potential of this new market and rapidly released products designed especially to appeal to the younger crowd.[139] The music industry was particularly quick to respond, especially as rock-and-roll artists like Elvis Presley proved to be sure marketing bets.[140] Records and merchandise sales for rock artists skyrocketed, and radio stations took greater interest in the new music form.

The music genre that helped American youth express their new attitudes was rock-and-roll. Their parents, however, had a different attitude about this new genre. With its origins in the blues music of black artists in the South, coupled with allegations that rock-and-roll led to teenage delinquency, many adults were suspicious of the new music. That made it all the more appealing to teenagers, who flocked to rock radio stations. The station managers at WGAC, however, were not about to change their format.

Radio as an industry was changing, but WGAC was not prepared to change with it. This was because WGAC's audience skewed older. Most of its listeners were those same people whom younger Americans thought could not be trusted—adults thirty years old and older. Because of the station's stronghold on that audience and its influence in the community, WGAC managers did not see a need to respond to the burgeoning youth market. The audience agreed. WGAC programming in the 1950s stuck with the tried and true: big-band sounds, news, radio dramas and comedies.

The one type of programming the station did not run, however, was rock-and-roll. That decision had a lot to do with how station managers understood their audience and also with the influence of on-the-air announcer George Fisher. Rock "was the music of the teenagers," said Cindy O'Brien, one of Fisher's daughters.[141] "It was just the times; the younger generation's music was usually in conflict with the parental generations' music." By the same token, when Fisher was managing WBIA, a station whose audience was younger and had different tastes, he had no problem with playing rock music. He even had Keith Beckum,

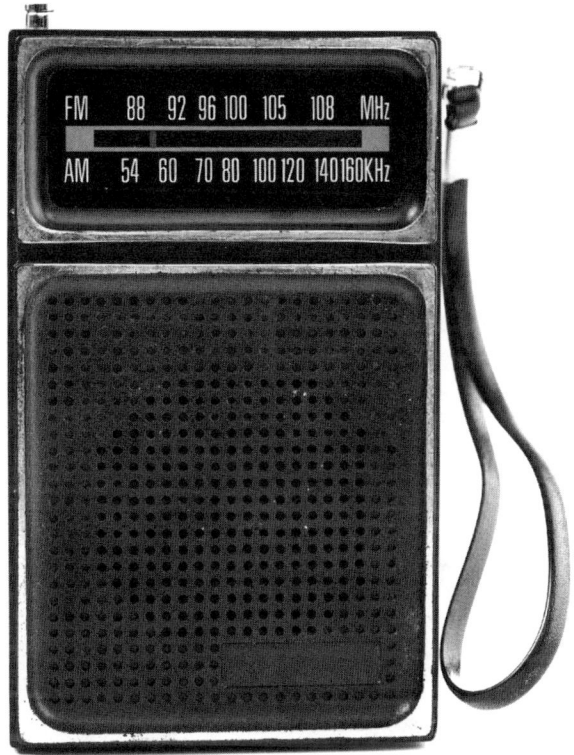

Transistor radios sold by the millions in the 1960s and 1970s as teens, and even an occasional taxidermist's assistant, used the new technology to take their music with them everywhere.

an announcer at the station, pull rock songs he liked to use in his morning show.[142] But during the 1950s and 1960s, when rock-and-roll was growing in popularity with the teenage crowd, he believed that it was not music that would appeal to the station's target audience.

"It's not like he hated rock-and-roll," said O'Brien. "He did play it, but his listeners were the ones buying the products he was advertising, and they were the people that were at least twenty-five to thirty years old and older. The younger people were not buying the products [that WGAC sold]."[143] So, instead of rock, Fisher would play the kind of music the station's older listeners wanted, which was primarily from the big-band era. His favorite artist was Frank Sinatra.[144]

Fisher, with his stilted cadence and strong-willed nature, was one of many who stamped his feet at the social changes that were influencing radio during the 1950s and 1960s.[145] His type of radio appealed to a

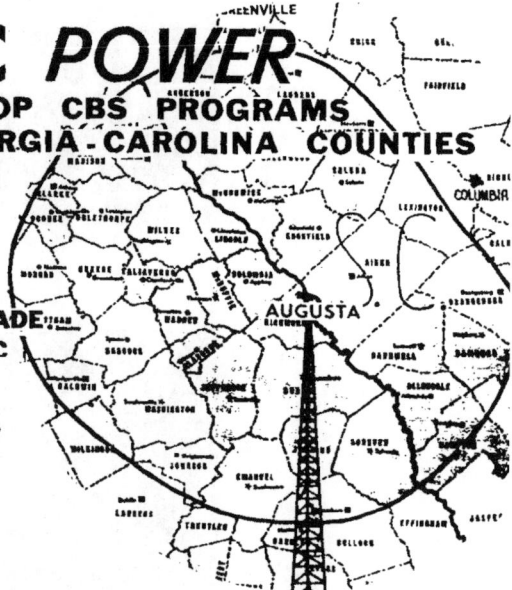

On the same day that CBS announced its new affiliation with WGAC, the station ran an ad in the *Augusta Chronicle* touting its new programming that featured Lucille Ball, Art Linkletter and Walter Cronkite. The ad ran on February 28, 1965.

more conservative crowd who wanted news, weather, community events and, of course, his trademark smooth, melodic tunes.[146]

Fisher's influence on WGAC in the early 1950s was profound but short-lived. He left the station within a year of joining it because of a programming dispute. Station owners decided to begin using some preprogrammed music rather than relying on show hosts to choose their own music. Fisher liked to pick out the discs he would spin on the air, and he liked being able to respond to requests from listeners.[147] Fisher was not willing to risk the connection he had forged with his audience through his music choices. Giving up his ability to piece together a program that would appeal to his regular listeners was asking him to turn his back on his audience. That was not in Fisher's plans. He wanted his listeners to think of him as a family member, or at least as a good friend.[148]

Perhaps one reason for the popularity of WGAC's programming in the 1950s and 1960s was that it gave listeners a place to go to get away from the social revolutions occurring around them. As the 1950s gave way to the 1960s, though, and the teen audience garnered more power in the radio marketplace, WGAC lost its dominant place in the Augusta market to competitor WBBQ.

The mastermind behind WBBQ's rise to dominance was George Weiss, a former WGAC employee who realized the opportunities that were opening up to stations willing to play rock. Thanks to new inventions like transistor radios, teenagers could, and did, take their music with them everywhere—beaches, bedrooms and backyards. Equally important, car radios made it possible for teens to listen to their favorite music while they were cruising down main streets and back roads. This was a market that Weiss wanted to capture when he made the decision in 1960 to change WBBQ's format to Top 40. He made the decision after watching New York City station WABC soar to the top of the charts after adopting a similar format.[149] Weiss made the right choice. By the late 1950s and early 1960s, popular Elvis tracks could be heard blaring just about anywhere teenagers gathered.[150] And in Augusta, this meant that the radio of choice for most listeners was no longer WGAC but WBBQ, which would hold the market's no. 1 spot for more than thirty-five years.

Because of its unwillingness to change with the times, the 1960s and 1970s were hard decades for WGAC. The station's hometown also faced hard times as the segregated South tried to grapple with the demands of African Americans for equal rights. In 1970, Augusta reeled from a race riot that tore through downtown, but this sad moment in the city's history proved to be a blessing in disguise for the city's former flagship radio station. WGAC's coverage of the Augusta riots emphasized the important role that radio news could play in informing the public and foreshadowed a successful path forward for the station. Buoyed by a strong devotion to the news format, the station was able to weather the period, and it would end up poised for a return to prominence.

WGAC's greatest technological problem in the 1960s and 1970s was the lower-quality sound available from the AM band. Rock-and-roll needed better sound quality than was available from that band. Listeners abandoned AM radio by the droves and migrated to the crisper-sounding FM band. As higher-quality radios became available that could really project the differences between AM and FM sound, audiences had an even better reason for tuning in to FM stations. Audiences waned for AM stations like WGAC.

Sound quality had become a big issue because rock music needed higher-fidelity reproduction, something that FM stations, which could broadcast at near studio-quality levels, excelled at. To provide this higher quality of sound to the masses of listeners, stations moved to the clearer FM channels. WGAC persisted on the AM band until 2011, when it began simulcasting on a sister FM station, as well as on its 580 AM frequency. The station struggled until it hit on a new format in the late 1960s: news. Having a more news-orientated station in town was beneficial when racial tensions flared in the early 1960s.

Kent Dunn, the general manager at WGAC since 1993, said that in addition to the emerging popularity of FM radio, the advent of stereo was another key component in WGAC's decline during this period. "Now, when you listen to a song, it's the difference between listening to an FM signal that's marginal and putting a CD in; it's the same kind of separation in quality," Dunn said. "I think that was really the key in what damaged AM. Now, you can listen to FM and your favorite music, like on WBBQ, and you don't have to hear all the static. The only radio format that is not affected by the AM static is the spoken word—news and sports." That is why so many AM stations have turned to news and commentary—the technology and the format are better matches. The change in format has been "kind of out of evolution more than anything else," Dunn said.[151]

The pocket-sized radio capitalized on this change in listening habits and the demand for popular music radio. A new invention, the transistor radio, catered to the demand for more music more of the time. The transistor radio offered both good sound quality and portability. This small, portable radio receiver, using transistor-based circuitry, was developed in 1954 and became the most popular electronic communication device in history. In 1961 alone, Americans bought some 23 million transistor radios, and billions would be manufactured and sold during the 1960s and 1970s. Teens and preteens—a category *Life* magazine referred to as "sub-teens" in 1961—bought "the little sets by the millions with their lunch money," but they were not the only market. Transistor radio manufacturer Channel Master Corporation, of Ellensville, New York, sponsored a contest in 1961 to find out where transistors were appearing. The largest single category was hospital and home delivery rooms. A California taxidermist's assistant also reported listening to a transistor radio set inside a mountain lion.[152]

Transistor technology entered the scene thanks to the work of Drs. John Bardeen, Walter H. Brattain and William B. Shockley at Bell Laboratories in the late 1940s.[153] The transistor functioned by conducting, modulating and amplifying electric signals. The transistor's durability and life expectancy, combined with its small size and affordability, allowed for the development of small radios, but at first the technology was devoted exclusively to military uses. It would be the mid-1950s before Texas Instruments would introduce the first commercial transistor radio, called the Regency.[154] The Regency could be purchased for $40 at the time, equivalent to just over $322 by today's standards.[155]

Transistor radios and the popularity of FM radio contributed to the decline in WGAC ratings. Harley Drew, who currently hosts a morning drive talk show on WGAC alongside News Director Mary Liz Nolan, was program director at WBBQ during these doldrums in the station's history. Drew recalled that WBBQ dominated local Augusta radio up through the 1990s while WGAC lagged far behind. WBBQ consistently pulled in a cume share of 30 or better, meaning that at least 30 percent of listeners in the Augusta area were listening to WBBQ. There were times when WBBQ did even better, pulling down at one peak point a cume share of 53.[156]

Scott Hudson, an investigative reporter at WGAC, said that the station got lost in the shuffle during these decades, taking a backseat to the decidedly more popular FM station WBBQ. "WGAC in its early years was basically the format that we have today, with the exception of throwing in some music and throwing some serials and that sort of thing, but it was basically information radio," Hudson said. "But during the '60s and '70s, WGAC just seemed to have dipped. It wasn't off the radar, but it certainly wasn't the success that WBBQ was."[157]

But on February 28, 1965, the radio station made a decision that helped it recapture some of the market share it had given up to WBBQ. Station managers affiliated the ABC station with the Columbia Broadcasting System's radio network. The vice-president of affiliate relations at CBS, William A. Schudt Jr., made the announcement. Operating on five thousand watts and at 580 on the radio dial, WGAC was in the strongest position on the radio. Jess Willard, president and general manager of WGAC, said, "We will continue [to make] good music [available] to our listeners…But some changes will be made in our dial format in order to bring CBS shows and features to listeners in our coverage area."[158]

The switch to CBS gave WGAC access to programming that featured Arthur Godfrey, Art Linkletter, Lucille Ball and Dear Abby on the entertainment side. On the news side, they gained reports from Walter Cronkite and Lowell Thomas.[159] Although Drew was not a part of the WGAC team during this time, he recalled that these changes were not as well considered as they might have been. The station's programming was varied, to say the least, Drew remembered:

> [WGAC] *was all over the place during that period. It wasn't focused on any one thing…You would get up in the morning and you would listen to a little "Masters of Music," I think is what they would call it. It would*

be anything from Bobby V. on up to Andre Kostelanetz. [Those DJs]
were trying to be announcers. It wasn't personality, it wasn't warm, it
wasn't fuzzy. Then they went into Don McNeal and the "Breakfast
Club" at 9 o'clock and then some soap operas, which were very big on
radio. And it was an AM station, so it had Paul Harvey.[160]

Despite its variety, or perhaps because of it, the station did not live up to its potential, Drew said. "It was a good station, it had a great signal, but it was boring," he elaborated. "People like to be talked to. They like to have a friend on the radio. And just to say, 'It's 10 o'clock, Cullum's the finest furrier in Augusta time, and we'll be back in a moment with more 'Masters in Music,' and here's 'ABC News'—that just wasn't it."[161]

Unfortunately WGAC was not able to capitalize on the advantages being a CBS affiliate offered. Drew explained that this was because the station lacked the kind of dynamic personalities that could hold listeners' attention. "WGAC…just didn't have any real personalities," he said. "Paul Harvey wasn't [even] a DJ, but he was the person everyone remembered about the radio station."

Another hindrance to WGAC, as well as every other radio station in the country, was the emergence of television as a dominant entertainment and news medium. By the 1950s, radio was beginning to be perceived as passé by many, and the industry was forced to adapt in order to remain relevant in the face of the tidal wave of momentum that television was gaining.[162] Authors Peter Fortale and Joshua E. Mills wrote that "[t]elevision had given radio a chill and a bad case of the shakes—but there was never any stoppage of vital signs. Station owners, investors, manufacturers, and all their employees had a vested interest in finding a new way to make radio work. What these people had to do was determine how to change, and then explain to their audiences that they were changing."[163] These authors claim that there were several ways in which local radio was able to survive and thrive. Chief among these was the presence of a popular disc jockey, "the announcer who played the records, read the advertisements, did promotion for the station, and sometimes read the news as well."[164]

While WGAC was struggling to stay competitive among all the technological advances and industry changes, Augusta was experiencing the worst of the civil rights era: a race riot that broke out in Augusta in the spring of 1970. Augusta was not immune to the racial tensions that were brewing in the South during this period, but it avoided the worst of the violence until that riot broke out in the streets of downtown on May 11. Six people died, and fifty-one arson fires were lit in six hours as Augusta blacks protested the death of a sixteen-year-old mentally handicapped boy while in the city jail.[165]

The 1970 riot was not Augusta's first encounter with civil rights demonstrators, but earlier protests had been peaceful. Students of Augusta's historically black college, Paine College, were particularly vocal during this period as they called for Augusta mayor Millard Beckum to desegregate Bell Auditorium in 1960 and as they advocated for equal treatment as customers of the city's mass transit system.[166] In May 1960, six students were arrested for refusing to move from the front of an Augusta bus when the driver asked them to do so. The students had been sitting next to or in front of the white passengers and were each fined forty-five dollars in Recorders Court.[167] In July 1963, Louis Harris, former editor of the *Augusta Chronicle*, wrote that "[the *Augusta Chronicle* sees] integration as being feasible in the field of education, voting, and job opportunities with either the city or county. In other words, we believe that the Negro should be given equality of opportunity in those areas in which he participates as a taxpayer."[168] Many Augustans felt otherwise.

Racial tensions culminated in Augusta on May 11, 1970, when riots broke out in downtown Augusta following the fatal beating of a sixteen-year-old African American inmate, Charles Oatman, at the Fourth Street jail at the hands of his fellow inmates.[169] Members of Augusta's African American community believed that the mentally handicapped youth had been neglected by the guards, and they incited a riot, starting at the Municipal Building before progressing down Broad Street and then across Ninth Street.[170] Whites who drove through predominately black neighborhoods were beaten and had their cars overturned. Fifty businesses in the city's center, many owned by Augusta's Chinese residents, were burned during the rioting. The Georgia National Guard was called to the scene, and by the end of the following day, the streets had been manned by more than two thousand troops.[171]

TIME SCHEDULE OF

Radio Addresses

by

Hugh Eugene Tudor

YOUR CANDIDATE FOR MAYOR

Save this announcement for future reference concerning these
political speeches over

wgac

Mon.	Sept.	3	8:45—9:00 P. M.
Tues.	Sept.	4	8:45—9:00 P. M.
Wed.	Sept.	5	7:45—8:00 P. M.
Thurs.	Sept.	6	9:45—10:00 P. M.
Fri.	Sept.	7	8:00—8:30 P. M.
Sat.	Sept.	8	7:35—7:50 A. M.
Sat.	Sept.	8	9:00—9:15 P. M.
Mon.	Sept.	10	8:45—9:00 P. M.
Tues.	Sept.	11	8:30—9:00 P. M.

TRUTH OR CONSEQUENCES
OF AUGUSTA'S FUTURE ELECTIONS!

A LEADER OF SATISFACTION—NOT PROMISES

WGAC has always been serious about its public service responsibilities. In 1951, not only was it airing radio addresses by mayoral candidates, it even bought advertising space in the *Augusta Chronicle* to announce the time schedule for those addresses. *From the* Augusta Chronicle, *September 3, 1951.*

Members of the Black Panthers arrived by the busload. Augusta historian Edward Cashin wrote in his book *History of Augusta* that "otherwise sensible people" were sitting "on their porches with loaded shotguns and to peer suspiciously out of their windows at passing strangers."[172] Though the riots received little real-time, on-the-spot live coverage by WGAC, the station was one of Augusta's best media for finding out about the riot because it was a news-oriented talk venue. The *Chronicle* and African American station WRDW were credited for much of the local coverage. In fact, singer James Brown (then WRDW owner) was credited for helping ease tensions following the riots that left six African Americans dead and more than sixty people injured.[173] The "Godfather of Soul," who was out of town at the time, returned home as soon as he received word of what was happening so he could help mediate. One *Augusta Chronicle* article estimated that out of about 300 people arrested, 150 were indicted for burglaries and destruction of property during the riots.[174]

Although these events were a part of life in the United States in 1970, the riot in Augusta, the "Garden City of the South," was the first major demonstration of the new decade and the largest riot of the period in Georgia. At the time of the riot, Augusta was home to more than seventy thousand people, and an estimated 53.6 percent (2006 estimate) of these residents were African American. The city was strictly segregated, with blacks living within the city's limits and whites living in the surrounding county and North Augusta. Slavery and Jim Crow are but memories of the past, but even today race relations continue to be contentious in city politics. With WGAC being a part of the CBS network, the riot was quickly catapulted into the national news headlines. WGAC kept the Augusta area informed with play-by-play reports of the riots.

One of the outcomes of the riots was that by the late 1970s, Augusta was looking for ways to address the city's racial problems. Public schools and other public institutions were working hard to desegregate, but the process of rebuilding sound relations between whites and blacks would continue on into the 1980s and 1990s and even into the new millennium.[175]

As Augusta worked to reestablish itself, so did WGAC, and the station jump-started that work by bringing in to WGAC two of Augusta's most beloved radio personalities: George Fisher and Pat Mulherin. Together, these two men would help end the station's dry spell and put WGAC back on the fast track to dominance in the radio market. Fisher returned to WGAC as vice-president and morning show host, and Mulherin came in as a sports broadcaster and top-flight advertising salesman.[176]

Chapter 5

MARCHING BACK TO SUCCESS WITH GEORGE FISHER AND PAT MULHERIN

Travis Highfield and David Lee

The 1970s and 1980s were challenging times in America. The Vietnam War was still raging in the early 1970s, and the withdrawal of troops began soon thereafter. The Watergate scandal gripped the nation as Richard Nixon resigned from office after being exposed for corruption. An energy crisis and the women's movement were also headlines throughout the period. Perhaps the greatest change, though, was cultural. Women and African Americans fought for equality, and college students protested to end the war. Drug use was at an all-time high—some 25 million Americans had experimented with illegal substances by 1979. New forms of music appeared that took hold of and changed the nature of American radio.[177]

Radio responded to these changes by adopting new formats more suited to the high-fidelity sound available from FM stations, but AM remained a strong component of the radio marketplace because of listener loyalty. AM listeners still tuned in to hear their favorite DJs spin the tunes on programs that had become a part of their daily routines. Competition between stations was fierce as each tried to field personalities that would draw more listeners. WGAC suffered in the battle for airwave supremacy at the hands of its cross-town competitor, WBBQ. WBBQ played the hip hits that appealed to teenagers. WGAC was not as quick to buy into a Top 40s or album rock format, and its ratings suffered. Station management needed to stem the loss of listeners and advertising, and it pinned its hopes on talent it managed to hire away from WBBQ: George Fisher.

Fisher had breathed new life into WGAC once before, but he had left the station to become program director and morning show host at WBBQ. A Chattanooga, Tennessee native, Fisher brought decades of broadcasting experience to both WBBQ and WGAC, but it wasn't all on-air experience. Fisher also knew music. A former high school trumpet player who also played in the Washington Redskins Band, Fisher had gotten his start in broadcasting as a musician. His first taste of radio came when he was a member of a thirty-two-piece U.S. Air Force orchestra. Each week, he performed with the orchestra on an NBC program known as *Contact*.

All those radio performances gave Fisher a taste for broadcasting. When World War II ended, the Columbus University graduate returned to school at the National Academy of Broadcasting. He did well enough in his studies that when he finished, he and several of his fellow students went in together to establish a radio station in North Adams, Massachusetts. Its call letters were WKOB.[178] Later, he returned to Washington, D.C., to work as a staff announcer at WMAL and WGMS. A few years later, he headed south.[179]

When Fisher left Washington in 1951, he was headed to Miami, Florida, to interview for a job. On his way, he heard about a program director position at WGAC. Former program director Warren Height had retired, and the station was looking for a replacement. Fisher interviewed for the position and was hired as morning talk show host.[180] He would become a fixture in Augusta radio, but like everyone else in the industry, his employer changed with some regularity. Fisher would establish two stations of his own: WAUG, the first station in Augusta to field a popular music show featuring big bands and other popular artists of the era, and WBIA (later).[181]

In between owning his own stations, Fisher helped build WBBQ into the most popular radio station in Augusta. He and several associates, including Pat Mulherin, developed the music-heavy format that made WBBQ Augusta's most popular radio station for several decades. As he later explained in an interview, the WBBQ schedule resembled a "Chinese menu" when he first went to work there. This was because its programming was trying to offer something for everyone. Fisher helped turn the station around and did such a good job at it that WBBQ came to dominate the Augusta market.

Fisher seemed always to need a challenge, though, and so after six years at WBBQ, he decided to try his hand at owning his own station again. In 1959, Fisher moved to Augusta's newest radio station, WBIA. He would spend the better part of two decades, punctuated by a four-year stint back at WGAC from 1972 to 1976, as WBIA's program director and morning show host. Mulherin, an associate from the WBBQ days, was with Fisher at WBIA for thirteen years. For four of those years, from 1976 to 1980, they would co-own the station.[182]

During his twenty years at WBIA, Fisher developed what would become his signature morning show. Not only was this show the most popular and listenable feature of WBIA, it would also become one of WGAC's most beloved programs after Fisher returned to that station in 1980.[183] The WBIA morning show ran for four hours. The station's biography of Fisher maintained, "From 6:00 till 10:00 a.m., Monday through Friday, thousands of radio dials are set at 1230 where George's friendly voice holds forth with a program that serves the public with news, weather, time checks, sports and community announcements interspersed between smooth, melodic music."[184] Fisher's morning show was distinct from the other disc jockey programs of the day because it appealed to everyone in Augusta rather than just a select audience. Fisher tailored his morning program so that it was suitable for everyone, of all ages and interests. He once told a newspaper reporter, "I don't ever want to say anything that will offend an 80-year-old grandmother or a 5-year-old child."[185]

Fisher would greet listeners each morning with his booming, "Good morning!" He believed it was his responsibility to get his listeners' morning off to a good start. He also tried to assure them that the rest of their day would be good, too, by reminding them to "Be good out there" at the end of his show.[186]

"He felt that it was his responsibility to always have a happy voice for his listeners early in the morning, and that meant he always kept a happy, he further realized that he did plan.

not have the skill and knowledge' Land that was best suited for
necessary to put Nature to work pasture, grazing crops, row crops.

AUGUSTA RADIO CLOCK

MONDAY
9-3-51

A.M. (WGAC 580	(A) WRDW 1480	(C) WBBQ 1340	(M) WJBF 1230 (N)
5:30 Ed Tacy	4 55 World News News at 5 55		
6:00 Ed Tacy	Dixie Early Birds Cousin Ed		News and
6:15 Farmer Chambers	and Farm News Cousin Ed		Gospel Parade
6:30 Hymn Singing	Dixie Early Birds Cousin Ed		News
6:45 Ed Tacy	and Farm News Cousin Ed		Hymn Sing
7:00 News Roundup	Dixie Early Birds Cousin Ed		Record Express
7:15 Hi Neighbor	and News Cousin Ed		Record Express
7:30 News & Hi Neighbor	Dixie Early Birds Cousin Ed		News
7:45 Hi Neighbor & News	and News Cousin Ed		Hugh Alison
8:00 Martin Agronsky	A News Roundup C White Rooster		Sammy Kaye
8:15 Musical Clock	Sing With Bing Cousin Ed		Guy Lombardo
8:30 News and Music	Renfro Valley C News Mkt Basket NBC News		
8:45 8:55 News	A Aug. News Cousin Ed		Ray Anthony
9:00 Breakfast Club	A Frank & Patti Daily Devotions		The Gospel
9:15	Frank & Patti Daily Strength		Train
9:30	Frank & Patti Baptist		The Gospel
9:45	Frank & Patti Hour		Train
10:00 My True Story	A Arthur Godfrey C Lyons		The Hillbilly
10:15 (10:25) Edw'd Arnold	Arthur Godfrey C Den		Hit Parade
10:30 Betty Crocker	A Arthur Godfrey C Cedric Foster		Red Foley N
10:45 Modern Romances	A Arthur Godfrey C Movie Music; Nws Hymn Time N		
11:00 Joyce Hayward	Arthur Godfrey C Ladies		Break the
11:15 Joyce	Arthur Godfrey C Fair News		M Bank
11:30 Hayward	Madelinn Maddox Queen for		Lorenzo Jones
11:45 Surprise Serenade	Platters & News a Day		M Dave Garroway N

P.M. WGAC 580	(A) WRDW 1480	(C) WBBQ 1340	(M) WJBF 1230 (N)
12:00 News	Wendy Warren Curt Massey Time Heath Talk Music		
12:15 Farm and Home	Stork Club; News Lomax. News		News
12:30 Hour with	Helen Trent Baseball		Hugh Alison
12:45 Claude Casey	Our Gal Sunday Chicago		Hugh Alison
1:00 Paul Harvey	A Big Sister C White Sox		Hugh Alison
1:15 Streetman	Ma Perkins C Cleveland		Hugh Alison
1:30 Baseball	Young Dr Malone Indians		Geo Hicks News N
1:45 Baseball	Guiding Light C Baseball		Dr Paul
2:00 Baseball	Our Singing Star Baseball		Woman in My Hou
2:15 Baseball	Perry Mason C Baseball		The 1230 Club
2:30 Baseball	Nora Drake C Baseball		Just Plain Bill N
2:45 Baseball	Brighter Day C Baseball		Frt Page Farrel N
3:00 Baseball	Hilltop House C Gospel		Life Can Be Beau
3:15 Baseball	News & Music Parade		Road of Life N
3:30 Baseball	Art Linkletter's Blues and		Pepper Young N
3:45	House Party C Boogie		Right to Hap'ness
4:00 Ed Tacy	Strike It Rich C Three		The Hugh
4:15 Presents	Strike It Rich C R's		Alison Show
4:30 Ed Tacy	Jam 'n Jive Three		The Hugh
4:45 Presents	Jam 'n Jive, News R's		Alison Show

MONDAY
9-3-51

P.M. WGAC 580	(A) WRDW 1480	(C) WBBQ 1340	(M) WJBF 1230 (N)
5:00 Ed Tacy	Jam 'n Jive Blues		Be Bop
5:15 Ed Tacy	Jam 'n Jive Boogie		Hour
5:30 Story Time	The Grab Bag Three		Be Bop
5:45 Jerry of the Circus	Curt Massey Time R's. Clubhouse		Hour
6:00 Six o'Clock News	News Roundup Cousin Ed		Be Bop
6:15 Claude Casey	Sports & News and the Folks		Hour
6:30 Sportscope	Eve With Hubbs Town Crier Nws Golden Trumpets		
6:45 Songs of Good Cheer	Eve With Hubbs Music by Benecke News		
7:00 (7:05) Hugh Grant	Eve With Hubbs Fulton Lewis		M News, Bing Crosby
7:15 Elmer Davis	A Jack Smith Show Sports Fare		Top Talent
7:30 Cisco	Club 15 Heatter News		M News of World N
7:45 Kid	Edward R Murrow Mutual Newsreel		1 Man's Family N
8:00 The Man from	Suspense Dance Time		Railroad
8:15 Homicide	A Box Scores		Hour N
8:30 H J Taylor	A Herb Shriner Club 13		Voice of
8:45 World News	A Talent Scouts Music; News		Firestone N
9:00 Jack	Radio Theater Music		Telephone Hour
9:15 The	Radio Theater to		Hour N
9:30 Jiver	Radio Theater Read		Band of
9:45 Jack	Radio Theater By		America N
10:00 News	Bob Hawk Show News, Starlight		Boston
10:15 Jack the Jiver	Sports Final Serenade		Pops Orch.
10:30 News of Tomorrow	A Augusta Ballroom Dance		Defense Bond
10:45 Dream Harbour	Augusta Ballroom Orchestra		Show
11:00 Musical	News & Analysis Baukhage Talking William Green		
11:15 Encores	Night U.N. Highlights Morgan Beatty N		
11:30 Musical	Owls Dance Surprise		
11:45 Encores	Club Orch. M Serenade		
12:00 Sign off	News, Night Dance Music News N		
12:05	Caller, News, til 1 Sign off Sign off		

WRTV CHANNEL NO. 2

By 1951, Augusta boasted four radio stations whose programming ranged from news to music to sports. *From the* Augusta Chronicle, *September 3, 1951.*

optimistic attitude," said his daughter Cindy O'Brien.[187] Fisher believed that projecting an optimistic outlook was an important step in appealing to listeners. He once advised a young co-worker to always smile when he said his name on the radio. "The audience can hear you if you're smiling," Fisher said.[188] To this day, that co-worker says he follows Fisher's advice. Whenever he does a news report, depending on the content and the image he wants to project into the listener's mind, he will either smile or have a serious face when he says his name. Fisher's happy attitude, the music he played, his daily devotionals, his emphasis on reporting good news and the constant laughter his show evoked were the mainstays the morning host used to get his listeners' days started off on the right track.[189]

Listeners loved Fisher. He forged close relationships with them, and one of his techniques for doing so was to talk about his family on the air. As a result, the morning show host had a loyal following. Even after his death, former listeners still talk about Fisher's programs. The fact that listeners still remember years later what Fisher did on his show demonstrates the influence he had on them and how greatly he affected their lives. They loved his down-home openness. He shared his family's secrets with his listeners and built a strong relationship with them that persists even more than a decade after his death.[190] His daughter Connie Bowles recalled overhearing a woman fondly describe how she used to listen to George Fisher and his "march the children off to school" segment when she was younger.[191] Fisher had such an impact on his listeners because he was dedicated to them, and that was true at each station where he worked.

When Fisher returned to WGAC after twenty years at WBIA, he brought Augusta's reigning morning drive-time show with him. He might have held the title of vice-president in his new job at WGAC, but his most important duty was his morning show. Fisher would continue that show until just before his death in the early 1990s, and his format never changed.

Every morning, listeners tuned in for a program of daily devotionals, discussion and the most popular feature of all: marching the children off to school. Each morning, as the school buses were running and harried parents were rushing to get the kids out the door, Fisher would recite, "Time to get 'em slicked and shined, buckled and bowed, zipped up, tuned up, tucked in, combed out. Look out. Be careful. Heeerrre they come marching out the door to school!"[192] The idea behind marching the children off to school came from Fisher's mother. "My mother, like most mothers, was caught up in all of that early morning confusion of getting children off to school," Fisher said in a question-and-answer interview his daughters found among the papers he left behind. "There were four of us (two brothers and a sister). She used to chide us about eating breakfast, brushing teeth, getting dressed, making up beds, and all the rest. Then she would say, 'Ok…ready or not, it's time for you to march out of here.'"[193]

People loved listening to him do the chant, and after years of doing it by himself, he started having children call in and march the children off to school. Fisher would also do small interviews with the children before they would march off to school.[194] "It was those type of things, the personal contact that he had with people, that made him so authentic," said Carol Spires, one of his daughters. "It made people want to advertise on his show because they knew that he was a fun person and they felt like he was talking to them."[195]

The chant became so popular that school-age children could recite it from memory. The chant became such a part of the fabric of Augusta's community life that Fisher would sometimes invite callers or studio guests to recite it in his place. Fisher's devotion to the children of Augusta earned him several awards during this period, as well as recognition from the then governor of Georgia, Joe Frank Harris. In his letter to Fisher, Harris remarked, "Through your broadcasts on WGAC radio, many children have been able to start their day with a bright outlook and a smile. It is rewarding to hear of individuals such as yourself who take the time to share their love with others."[196] One *Augusta Chronicle* article noted, "The start of the school day wasn't complete for many families without listening to Mr. Fisher's daily conversation with a child."[197] In 1986, Fisher was the recipient of the Golden Apple Award, an award annually presented to reporters for exceptional coverage of education activities.

Some argue that Fisher's ability to share "love with others" is what made him immensely popular in the first place. His informality while on the air made listeners feel more like family members than acquaintances.

WBIA's headquarters in the 1960s. *Photo by John Diskes.*

His mellow voice and demeanor made it seem as if he were in the room along with his listeners. Keith Beckum, then a producer and copywriter at Beasley Broadcasting group, WGAC's parent company, was once quoted as saying, "He gave you a sense of authenticity."[198]

Fisher could put his audience at ease with his cordiality, and it made him feel just like a neighbor.[199] As Fisher once said in an interview, "I think the man on the mic should never forget that he is a human being talking to other human beings."[200] He was so well liked by the city of Augusta that former mayor Lewis Newman declared June 17, 1975, as "George Fisher Appreciation Day."[201] Fisher later went on to earn the first West Augusta Rotary Louis Harris Award for a radio personality in 1984, an award named after the late executive editor of the *Augusta Chronicle* that is presented annually to members of Augusta media who demonstrate excellence in their work.[202]

Fisher's popularity with his listeners gave him the sort of influence that advertisers wanted to tap into, but his endorsements were not for sale to just anybody. Fisher would only endorse products in which he believed.[203]

This connection with his audience is what garnered Fisher a loyal fan following and what made him successful as a salesman. Station fans knew that if he said a product met his standards, it was a reliable product.[204] Fisher's strong devotion to advertising products that he believed in was also part of his work ethic.

According to his daughter Cindy, "His work ethic was impeccable; [he was] extremely devoted to always doing the right thing no matter the situation. Honesty and integrity were the cornerstone[s] of his beliefs. He was known to refuse advertisers if he felt that they or their product were not something he could stand behind. His work ethic was strongly rooted in making sure everything he did was a good thing for him, his advertisers [and] his family."[205]

Besides having a strong work ethic, Fisher was very detail-oriented and organized, which meant that he liked to have everything in its place, both at work and at home. He was often found in the control room of the radio station because he had to make sure that he could grab whatever he needed at a moment's notice, whether it was commercial tapes, records or news reports.[206] His organization was a part of his routine and would stick with him throughout his life. Scott Hudson said that when he worked with Fisher later in his career, he also discovered how detail-oriented Fisher was.[207] Fisher had a box with folders that had the commercial copy he would read live on the air. Hudson said he would pull all the commercials and use carts, which would play the commercials and then rewind themselves automatically to the beginning.[208]

When he first began working at the station, Hudson was responsible for putting Fisher's advertising carts in the proper order. "God forbid I made a mistake and got the wrong copy somewhere," Hudson recalled. "He had a photographic memory, [and] the next morning he would come in and go, 'Scott, Milton Reuben Chevrolet. What do they sell? Chevrolets? Pontiac Master. What do they sell? Is there a difference between the two?' 'Yes, sir.' 'Well you put Pontiac Master in the six o'clock hour when it was supposed to be Milton Reuben, and I expect you to pay a little more attention next time.'" That was George, Hudson explained. He was exacting in his standards, and even one misplaced commercial was worth making a fuss over.[209] All this hard work and attention to detail made

Fisher good at what he did and kept the attention of his listeners for the many years he was on the radio.

Despite Fisher's many accomplishments and accolades, he could not bring WGAC back into prominence by himself. For that, he needed the help of his longtime business partner, Pat Mulherin. Like Fisher, Mulherin fell into his radio career by accident. His initial life plans never included seeking a career in broadcasting. According to Mulherin's wife, Elizabeth, he did not know that he was going into the radio business until he got his first job. Mulherin finished college at Spring Hill College in Mobile, Alabama, in 1943. He graduated a year early so he could enlist in the U.S. Army, but by then, World War II was over. Mulherin's uncle got him his first job at Augusta radio station WRDW.[210] Mulherin moved on to another Augusta station, WGUS, where he held the position of vice-president and general manager of the station. He also had an on-air slot during which he played country and western music and did news and sports. Like Fisher, Mulherin also moved between radio ventures. Before landing at WGAC in the 1980s with Fisher, Mulherin also worked at WBIA and WFNL.

The stations where Mulherin really made a difference, though, were WBBQ and WGAC. Mulherin helped put both these stations on the map. At WBBQ, Mulherin became a popular sports broadcaster, thanks to the training he had received from Thurston Bennett. Bennett was already a popular sportscaster and took Mulherin under his wing and trained him in the art of reporting sports on the radio, including play-by-play action. Mulherin first started announcing sports with the Augusta Tigers, a minor-league baseball club, but he soon moved into the booth to do play-by-play on WBBQ, and he also did some television sports for Augusta Channels 6 and 26. However, radio was his real love. As a sports broadcaster, Mulherin got to witness the talents of players such as Hank Aaron and Joe Torre as they passed through Augusta during their minor-league stints.[211]

Mulherin did not travel with the Tigers on the road, but that did not stop him from doing play-by-play broadcasts. To give his shows a "you are there" feeling, Mulherin used a gadget that Bennett had developed that consisted of a ball and spring that made a clicking sound. Over the radio, the gadget provided the sound effect of a baseball hitting a bat. He would

WBBQ's headquarters in the 1960s. *Photo by John Diskes.*

combine that gadget with play-by-play wires off Western Union to give listeners the sense that he was on the scene calling the game. When the wire said a player got a hit, Mulherin would use the gadget to add a sound effect.

"Nobody knew Pat wasn't at the game," Elizabeth Mulherin said. "They just believed he was there doing the play-by-play."[212] Mulherin was also forced to re-create action from a Masters City Little League team's state championship game in Columbus, Georgia, when the telephone company failed to properly set up the equipment necessary to broadcast. When he was finally able to talk on air, he did play-by-play as if the game was happening live in front of him when, in fact, he was relaying previous information.[213]

Mulherin also did play-by-play for various schools in the Augusta area, including the Academy of Richmond County and Aquinas and Evans High Schools. His favorite school, though, was Augusta College, today Augusta State University. "He loved Augusta College," his wife said. "He was the starter for the golf team when they had their tournaments here. Marvin Vanover was the coach of the [men's] basketball team, and he always said Pat was like his second coach. He would tell him what he should have done. Marvin would call him up and tell him to meet at Shoney's to talk about the games."[214] Augusta College recognized Mulherin's influence on its athletics after his death. In 1988, the school established a scholarship in his honor that was celebrated during a halftime commemoration at a basketball game. The 1989 Augusta State basketball media guide was also printed in his honor.[215]

Besides sports casting, Mulherin provided the nonbroadcast afternoon drive-time bookend to complete Augusta's working day with his *Going Home Show*. The "show" was performed each afternoon as Mulherin sat outside on the station's lawn and played music while greeting passersby on the way home from work that day. As someone who knew everybody, this was exactly Mulherin's cup of tea, according to his wife. "Someone would greet Pat and say, 'Don't say you see me. Nobody is supposed to know I left work early,'" she said.[216]

Mulherin brought his popularity as a sportscaster with him when he and Fisher came to WGAC in the 1980s. His fans from WBBQ followed him over to the new station. Perhaps his most popular work at WGAC was his football scoreboard show each Saturday afternoon following the University of Georgia game. Sponsored by various companies over the years, the call-in show provided updated scores from that day's college football action, as well as the opportunity for listeners to call and request scores, which Mulherin dutifully announced. "The Saturday scoreboard was extremely popular, and to this day we miss it," said Bernard Mulherin, Pat's brother, who is a senior superior court judge in Augusta. "You can't get scores that way anymore, and back then, all it took was calling in to Pat Mulherin, and he would always answer the phone. Mrs. Cooney would call in about [the University of] Notre Dame. He had a Clemson [University] guy who would call in, different ones like that."[217]

Mulherin also played a major role in WGAC's Masters coverage each year. The station was home to Masters reports in Augusta, providing updates every twenty minutes during the tournament. Doing the updates earned Mulherin the opportunity to interview some of golf's greats of the period, such as Arnold Palmer, Jack Nicklaus and Gary Player.[218]

Because they were so well liked, Fisher and Mulherin excelled as businessmen. Some considered the two a "dynamic duo" given their ability to advertise and sell products on the air.[219] Mulherin could sell the ads like nobody else, and Fisher had the prestige to sell products with his personal endorsements. Listeners trusted Fisher. They knew that he would not lie to them about sponsors' products. An endorsement "coming from George, that meant something," Bernard Mulherin said.[220] For that reason, Fisher only endorsed products that he trusted and believed to be fit for Augustans. In one WBBQ broadcast, Fisher remarked on his morning program that he was "particularly proud of [their] sponsors. They are a pretty selective group. Almost all are local business people and our real personal friends."[221] Advertising spots on Fisher's morning show were so popular with advertisers that the revenue from that show could pay for the rest of the station's programming. Broadcast partner Matt Stovall was quoted as saying, "If George Fisher said it, it was the gospel."[222] Keith Beckum, who worked with Fisher at both WBIA and WGAC, recalled a story told to him by an Augusta furniture salesman about the power of Fisher's endorsements. One day, a woman came in and asked the salesman to show her the recliners that Fisher had advertised. The salesman showed her the chairs, and the woman decided to buy one, commenting, "If George Fisher says it's a good chair, then I'll take it." The salesman reminded the woman that the store paid Fisher to sell their chairs, but the woman was unimpressed. George Fisher would never endorse a shoddy product, she assured the salesman. If he said it was a good chair, it was a good chair.[223]

Mulherin's wife described her husband as "a happy Irishman" who loved his family, radio and barbershop quartet. He died while performing at what was then Augusta College, another of his loves. During a show at Augusta State in May 1988, Mulherin's barbershop quartet was performing an act about doctors and a patient. Mulherin had the role of the patient. "He started having an attack," Bernard Mulherin remembered, "and John Fisher, the doctor, thought he was just doing a darn good job of acting. Finally, they realized he wasn't acting; this is real." Mulherin died of a heart attack at the age of sixty-four.[224] He left behind a legacy of radio work well known to Augusta's residents. His involvement in sports, handed down by Bennett, became the new standard for reporters in the market. But his personality and community involvement, volunteering as a high school football coach and helping others in need, also made an impact on Augusta. The passing of Mulherin left Fisher to generate the station revenues without Mulherin's sales abilities.

Fisher, though well liked by Augustans, was also known for being a little headstrong about what he believed in, and that included radio formats. Even as listener tastes changed, Fisher resisted alterations in his format. He preferred to stick with the brand of radio that he knew best. Consequently, when Station Manager Kent Dunn and Program Director Harley Drew decided it was time for Fisher's daily devotionals to end, he looked them in the face and proclaimed, "You're all going to hell!"[225]

The decision to do away with the morning devotionals was not because managers wanted to distance the station from its Christian roots. WGAC still broadcasts Sunday morning services. However, station research had demonstrated that the devotionals were not meeting listener demands. As Dunn said, "We just couldn't just expect everyone in Augusta to be Christians, and we wouldn't want to alienate any prospective listeners."[226]

By the time the 1990s rolled around, Fisher was declared Augusta's favorite radio personality in a survey conducted by the *Augusta Chronicle*.[227] Fisher worked hard to earn that accolade. He arrived at the station at 5:15 a.m. to prepare for his three-hour morning show that ran from 6:00 a.m. to 9:00 a.m., Monday through Friday. When he wasn't on air, he was working to tape radio spots that required his voice and mull over the sales department side of his job. Fisher did not work all the time, however. Like Mulherin, Fisher had a life outside the radio station.

In addition to his duties at WGAC, Fisher participated in community activities. He played with the Augusta Concert Band performances and stayed busy with speaking engagements at schools, churches and civic organizations. By 1992, Fisher had served on the boards of directors for the Augusta, Richmond County Library, the Salvation Army, the Augusta Players, the Augusta Opera Association, the Golden Harvest Food Bank and the Easter Seal Society, just to name a few.[228] Fisher held a Gold Card lifetime membership with the Georgia Parent Teacher

Association and was named honorary chairman of the Georgia Council of PTA in 1989.[229]

Fisher lost his battle with cancer on July 9, 1995, at the age of seventy-three. He had finished his last broadcast from the deck of his house only a few weeks before. He left behind his wife, four daughters and a powerful legacy, not to mention shoes too huge to fill. On July 10, 1995, WGAC's morning program featured a tribute to Fisher that included a taped montage of his career highlights. Listeners were asked to call in to share their memories of Fisher from prior years.[230] He was posthumously inducted into the Georgia Radio Hall of Fame in 2008 for his legacy in the Augusta radio market.

Chapter 6

FROM INSOMNIAC RADIO BACK TO THE TOP

John–Michael Garner and Kristin Hawkins

Through the 1970s and early 1980s, AM radio was moving toward the doldrums. FM had caught up to AM in terms of audience share, and by the late 1980s, it had attracted 85 percent of the American radio audience—despite the fact that there were only about one thousand more AM stations in America than FM stations. Given that music had migrated to the FM band due to its better sound reproduction, AM started eyeing a different direction: talk radio.[231]

Talk radio in the late 1970s and early 1980s was called "Insomniac Radio." The joke was that between about midnight and 6:00 a.m., only insomniacs were up listening to the radio. Nationally syndicated shows with personalities such as Larry King, Tom Snyder, Sally Jessy Raphael and Dr. Joy Browne filled AM airwaves during the nighttime hours, and their audiences were small.

A confluence of events in the late 1980s changed all that, though. AM radio was beginning to consider the advantages of news-talk programming since sound was less of an issue for that genre. However, talk radio had some potential drawbacks. The Fairness Doctrine required broadcast outlets to represent both sides of any political discussion. Stations that violated the Fairness Doctrine faced fines or the potential loss of their licenses, so politics was out for most talk shows. Show topics had to be innocuous. Larry King interviewed celebrities, for example, and Joy Browne gave advice about relationships.

In the late 1980s, however, the FCC under President Ronald Reagan began looking at deregulating broadcasting, and one of the measures that came to their attention was the Fairness Doctrine. The Fairness Doctrine had been adopted originally during a time when only the AM radio band existed, so everyone who wanted a radio station could not have one. Because of the scarcity of radio frequencies, the FCC determined that those who had custody of a frequency should operate it in the public interest, convenience and necessity. This meant that they should not allow any sort of bias to creep into their news content, which then meant that broadcasters were required to represent all political perspectives on any public issue they covered. By the late 1980s, though, the original justifications for the Fairness Doctrine were gone. The FM band existed, and it offered a multitude of frequencies. Further, there were new media, such as cable television, that were not subject to FCC regulations. Over-the-air broadcasters were stymied by the Fairness Doctrine in ways their nonbroadcast competitors were not.[232]

In 1987, the FCC got its chance to act on its Fairness Doctrine concerns. Syracuse, New York radio station WTVH, owned by Meredith Corporation, had broadcast several ads that supported the construction of a nuclear power plant. The Syracuse Peace Council requested equal time under the Fairness Doctrine to state its case. The FCC had been studying the Fairness Doctrine and found that it did not actually promote the public interest. Members believed that it should not be enforced any longer. The Washington, D.C. District Court, which heard the case, agreed with the FCC and gave the agency permission to drop the Fairness Doctrine. Congress objected and passed legislation to retain the Fairness Doctrine, but President Reagan vetoed the legislation, and there were not enough votes in Congress to override the veto.[233] With the demise of the Fairness Doctrine, the door was opened to a new kind of political talk show and a new role for AM radio.

It was a match made in heaven. In 1985, roughly 100 radio stations were using news-talk formats. Within five years of the repeal of the

For about twenty years, Mary Liz Nolan has been CSRA's reliable WGAC news radio host. *Photo by Patricia Johnson of Saturated Simplicity Photography.*

Fairness Doctrine, there were 500. Three years after that, in 1998, there were 1,350.[234]

WGAC was one of those American radio stations whose future success was ensured when it adopted the news-talk format. The station had been floundering for some time in the 1980s. Its programming lacked structure, and its advertising, outside of its popular morning drive program, was falling off.[235] WGAC still had its powerful signal, but it lacked an identity in the community. It had a large broadcast radius, and that gave the station potential, but it just was not going anywhere.[236]

The station's owner in the late 1980s was Bob Beckham. Beckham was an Augustan and an astute businessman, but he did not have any radio experience. He knew that the station needed to beef up its news operation, but with the station not performing at top speed, money for salaries was at a premium, so Beckham decided to take a chance on a new addition to the station's news team. In late 1987, he hired a young local by the name of Austin Rhodes. Rhodes would eventually become a cornerstone of WGAC's comeback as Augusta's premier news-talk

For more than fifty years, Harley Drew has been a jovial Augusta broadcaster. He remains as host of *Augusta's Morning News*. *Photo by Patricia Johnson of Saturated Simplicity Photography.*

station. WGAC's news director, a seasoned journalist named Greg Patin, took Rhodes in hand and taught him how to be a newsman.[237]

Just a few months later, a national talk radio show would premier that would change the face of radio: *The Rush Limbaugh Show*. Limbaugh was the first of his kind. He was witty, controversial and he gave the common person access to a national forum. Limbaugh's show would revitalize national AM radio like nothing before. He demonstrated that nationally syndicated radio could still deliver a mass audience. By 1992, radio was again a national medium that could create national stars. Today, the combined audience for national daytime radio is on par with, and sometimes surpasses, the audience for national television networks.[238]

But when Limbaugh's show first aired, no one knew how successful he would be. There was much debate about where the format was going. Not only was his style completely different, but it had also been fifty years since there had been much of an audience for radio in the middle of the day.

People were working, and critics did not see how Limbaugh's show could gain an audience, but he made it work.[239] Limbaugh knew how to keep his audience and gain more listeners. His secret was to keep controversy and dispute as the main points of his show. It was not long before critics and even political figures realized how influential Limbaugh's words actually were. No one before Limbaugh could persuade an audience's opinion like he could.[240] But Limbaugh had another secret, too, and that was that his show gave voice to those who had previously felt voiceless. Alan Colmes, one of the few liberals to succeed as a talk show host, made this point in a discussion about why talk radio worked: "Talk radio gives voice to those who, until its advent in the popular culture, felt they had no voice." Another noted talk show host, Blanquita Cullum, agreed with Colmes. She argued that the value of talk radio is the opportunity it gives citizens to "talk back. It's all about our freedom of speech." Talk radio is appealing, Cullum argued, because it is interactive. "The host needs the audience; the audience needs the host."[241]

It has not hurt that Limbaugh, like so many other radio talk show hosts, is conservative and thus good at attracting a very specific demographic—specifically, a conservative, wealthy and mostly Republican audience. He plays well to the more conservative component of Augusta, and from a business sense, that is a very attractive group of people for advertisers to reach.[242]

Within only a few years of his national radio debut, *The Rush Limbaugh Show* had been picked up by hundreds of AM stations across the country, and millions of listeners, who refer to themselves as "dittoheads" were listening to him each day. Limbaugh helped save the dying AM band and pave paths for other controversial radio programming across the country. However, at least in the early days of his show, Limbaugh could do nothing for WGAC, for he did not run on the station, which was making so little money that then owner Bob Beckham was forced to return the struggling station to Beasley Broadcasting, a radio conglomerate out of Naples, Florida, in 1993.[243]

The irony of Limbaugh's introduction to Augusta was that the man who held the rights to *The Rush Limbaugh Show* was Carol Red. Red owned an urban-style radio station named Foxy 103. The urban format and

Mary Liz Nolan and Harley Drew wake up Augusta listeners with the daily on-air show *Augusta's Morning News. Photo by Patricia Johnson of Saturated Simplicity Photography.*

Limbaugh's show were essentially antithetical to each other. Still, Red tried to hang on to the show, but he kept losing audience when the show ran, so he eventually gave it up. WGAC stepped in to buy the rights.

Having the rights to the Limbaugh show, though, did not mean automatic success. WGAC's programming was still very fragmented. An average day for WGAC in the late 1980s consisted of George Fisher in the morning, elevator music, Rush Limbaugh, more elevator music, *Voice of the People*, more elevator music, Dr. Joy Browne and then more elevator music until Tom Snyder came on. The station lacked a consistent market or direction. It really adhered to no true format.[244]

But the "Rush bandwagon" gave the station a starting point. Managers jumped in to see if they could pull any money or listeners from his show. Limbaugh's show had a specific audience. His listeners were, at the very least, mostly politically aware. They were generally educated on the issues he talked about.[245]

It took WGAC a while, but eventually Limbaugh's success started to rub off on the station. "Talk radio blossomed into reality for us and for a lot of other stations with the arrival of Rush Limbaugh," said Harley Drew, a morning talk show host for WGAC. "He was the one that had the magic spark that made people want to listen every day."[246]

The real success came when station managers decided to anchor the Limbaugh show with a local talk show that would air during the lucrative afternoon drive time. The host of that show was Austin Rhodes, the young newsman who would become the acknowledged king of Augusta talk radio. Rhodes had left the station for a brief stint with television news, but he returned to WGAC in 1992 as host of the station's afternoon talk show, *Voice of the People*. At the time, few could have been able to foresee the impact that he would have on the Augusta area.

The often caustic and irreverent Rhodes quickly became recognized for his take-no-prisoners approach to talk radio. Unafraid to tackle the tough issues and call powerful local and national leaders on the carpet, Rhodes has functioned as something of a lightning rod in Augusta radio. During his nineteen-year tenure at WGAC, he has established a loyal base of listeners. About as many love to hate him as love him. Whatever his critics might think of Rhodes, there is one thing they cannot deny: they listen to him.

From an early age, Rhodes knew that he wanted to be involved with radio. As a child, he listened to the Atlanta Braves on WGAC with both of his grandfathers. As a youth, in the mid-1970s, he began to listen to the show that he would host almost two decades later. That show was *Voice of the People*, and it first aired during the 1976 United States presidential campaign. Jimmy Carter, a former Georgia governor, was seeking the nation's highest office, so interest in the campaign was high in Augusta. Rhodes had a grandfather who was a political junkie and who would keep the radio on WGAC's afternoon talk show (*Voice*). Barry Yon, who used the on-air pseudonym of "Bob Young," hosted the show. Although Yon never pulled in huge ratings, Rhodes was a devoted fan and called him "probably the best radio talk show host that this area has ever seen."[247]

Rhodes said that he called in to *Voice of the People* as often as twice a week, which earned him the nickname of the "WGAC Whiz Kid" from Yon. Rhodes would not meet Yon in person until years later, but even as a youngster, he was impressed by the level of mastery that Yon had over the microphone. Yon made being a radio host "sound like one of the coolest jobs in the world," Rhodes said. He was willing to do "just about whatever it took" to get his foot in the door.[248]

Once Rhodes was in high school, he volunteered to help out with whatever the station needed. That landed him the opportunity to help with Friday night broadcasts of high school football games. His partner in those broadcasts was one of Yon's occasional fill-ins as host on *Voice of the People*, Matt Stovall. Rhodes said that the experience of working alongside Stovall and "watching and learning how he did things" was "extremely valuable."[249] During his first year of college, Stovall "hired" Rhodes as an unpaid assistant, a position that he held on to "for a number of years," simply to "be close to the broadcasting industry."[250] By his second year of college, Rhodes was working part-time on the weekends at the station, in addition to his work with Stovall on high school football.

Rhodes did eventually get hired at WGAC, but that was at the height of the station's doldrums, and he was not optimistic about the station's direction. Rhodes said that WGAC's inability to keep up with changing trends in the radio industry influenced his decision to leave. "I kind of saw the handwriting on the wall that the talk radio format had not really taken hold here the way it had in other markets, and WGAC was fading fast," Rhodes said. "I wanted to get in television news anyway."[251]

In early 1989, Rhodes decided to try his hand at television news with WRDW-TV. He would work at that station for the next three years. In 1992, Rhodes was lured back to radio, and when he returned to WGAC, he found that the station had been stripped down to the bare essentials. When he had left for television three years earlier, WGAC's news department had three members. When he returned, it had no one except him.[252]

Rhodes said that his directive when he was hired for his second go-round with WGAC was to improve the quality of the station's local coverage. WGAC had the Limbaugh franchise, which played an instrumental role in helping revive the network, but there was little emphasis on local news. The then host of *Voice of the People*, Keith Beckum, was a "great guy," but Rhodes said that he was not as well suited to the program as Yon had been. After Rhodes was hired to do news, local news and information became more of a priority for the station, according to Rhodes. That

Austin Rhodes is indisputably Augusta's king of radio in the new millennium, but like most of WGAC's personalities over the years, he is also active in the community. One of his involvements is with local theater companies. Here he is wearing his costume for an upcoming play.

decision, coupled with the appeal of the Limbaugh show, ensured that the station's success would be "like a snowball going down a hill after that."[253]

Scott Hudson, an investigative reporter at WGAC, said that Rhodes's local flair added important variety to the station's talk show block. He explained that Rhodes generally does not get into national politics whatsoever. The talk show host figured, "Why, if we've had Neal Boortz and Rush Limbaugh on for six hours, should I regurgitate what they've done for six hours?" The basis of Rhodes's success, in Hudson's estimation, has been his ability to get people really involved in understanding events and issues in the local area. Broadcast coverage, especially of local politics, was scant before Rhodes's show, Hudson said, and even the *Chronicle* might not always have devoted as much attention to local matters. "But with Austin, he's going to tell you everything," Hudson said.[254]

Like Limbaugh, Rhodes is a highly polarizing figure, yet his aggressiveness and entertaining style have made him one of Augusta's most successful talk show hosts of all time.[255] Rhodes said that he attributes

The *Columbia County News Times* ran this cartoon as a testament to the prowess of WGAC's news gathering. *J.B. Woody,* Columbia County News Times.

"100 percent of his show's early success to the presence of Limbaugh on the station…Limbaugh just created this opportunity for the local person to step in and say something, and if you knew what you were saying, and you knew what you were doing, then it worked."[256]

Aside from his knowledge of the local community, Rhodes said that his willingness to speak out about issues that other newsmen were afraid to touch was another reason why his show caught on quickly.[257] Mary Liz Nolan, WGAC's news director, said that Rhodes's dogged pursuit of a story, no matter how difficult it might be, is what makes him such a valuable resource to the station. "Where there's a will there's a way," Nolan said. "He's going to do whatever it takes to find the information he needs."[258]

Despite his unapologetic style, Rhodes said that he has never lost sleep worrying about retribution as a result of his bombastic commentary. "People are big, fat scaredy-cats when it comes to discussing certain things, but I can count on one hand the number of radio talk show hosts that have been violently attacked or killed, and usually it has to do with either foreign people with a political agenda or just a lunatic that would kill anybody," he said. "I've never been afraid for my safety."[259]

That's not to say that Rhodes has never had his safety threatened by angry listeners, however. Rhodes said that he's had "a couple of people locked up" for making threats against him—though, to his knowledge, the perpetrators were threatening him because they were "lunatics" and not because they were offended by anything he had said.[260] One man, a frequent caller to Rhodes's show, served two years in a federal penitentiary after sending detailed death threats to Rhodes's home address and calling Rhodes at his home. When authorities detained the man at his house, they found that he had more than eighty pictures of Rhodes, which he had presumably clipped out of newspapers, nailed to a wall in his apartment. Rhodes said that the man, a failed radio talk show host himself, had a history of attacking broadcasters, which Rhodes believes was rooted in intense jealousy. Rhodes said that he accepts incidents such as these as part of his job description. "These things happen when you're a public figure," Rhodes said. "Television news anchors get those kinds of threats all the time. It happens, and you just learn to deal with it."[261]

Keith Beckum has worked at WGAC since the early 1980s. He got his start in radio with George Fisher at WBIA. *Photo by Debbie van Tuyll.*

Although Rhodes might have been aware that his show was thriving in its early days, it was not until a fall afternoon during the 1992 U.S. presidential campaign season that he realized that his voice was being heard by some of the country's most powerful leaders. Thanks to the success of Limbaugh, the Republican Party during that time was beginning to embrace talk radio. When Republican leaders would visit a city, they would often seek out the chief talk show host in the area.[262] In October 1992, just three months after Rhodes had taken over as host of *Voice of the People*, then vice president Dan Quayle came to Augusta for a campaign rally at the Augusta Mall. During this time, many conservatives had taken to calling Democratic presidential nominee Bill Clinton a "waffler" on the issues, and Rhodes suggested during his show that everyone attending the rally should bring waffles with them.

Rhodes estimated that roughly five thousand showed up with some form of waffle or waffle-related items in hand. It was "the first time that a crowd had manifested with waffles in hand."[263]

During this time, Rhodes was not only hosting *Voice of the People* but also doing a news and information hour. Because of his continued involvement with the station's new programming, he was able to get a press pass to attend the Quayle rally. At the rally, a man in a dark suit wearing dark sunglasses approached him. The man, who turned out to be a Secret Service agent, ordered Rhodes to gather his belongings and follow him. A perplexed Rhodes tried to find out from the mysterious man where they were headed, but to no avail. Finally, the Secret Service agent told Rhodes that William Kristol, the vice president's chief of staff, wanted to speak with him. Upon meeting Kristol, Rhodes informed him that he was the person who instigated the waffle outbreak, to which Kristol replied, "That was pretty effin' funny."[264] Kristol then told Rhodes that they had to leave for the airport within the next ten minutes and asked if Rhodes wished to speak to Vice President Quayle. Of course the young talk show host gave an enthusiastic yes.

Rhodes said that he spoke to Quayle for several minutes about a variety of subjects as they drove to the airport at breakneck speed. "Quayle makes this little speech, jumps in the limo and we go peeling out of the parking lot of the Augusta Mall going 90 miles an hour," Rhodes recalled. "Politicians, the high-ranking ones, love to drive fast. And they will literally, for the president and vice president, block off traffic," Rhodes recalled. "So we did 110 down the Bobby Jones Expressway going back to the airport where Air Force Two was on the tarmac." Rhodes said that this was the first time his show had made a big splash. "People were a little bit surprised that the vice president or the president or anybody at the national level would embrace talk radio so obviously at the expense of other media," he said. "But they recognized us at that point as advocacy media, and they went where the support was. So that kind of put the show on the map."

It was shortly after his encounter with the vice president that station managers asked Rhodes to revise the *Voice of the People* into his very own *Austin Rhodes Show*, complete with multiple versions of his own jingle. Almost twenty years later, Rhodes and his show are still having a substantial influence in the local area. Rhodes believes that because he has "been around for so long, and because I grew up in the area, I've got a little experience that's going to be hard to duplicate unless you can bring somebody in that's done exactly what I've done their entire life and lived it all. There are a few people like that; thank God none of them are in talk radio. But I really think I addressed a lot of people and a lot of issues that folks were afraid to talk about."[265]

In 1995, with Rhodes firmly entrenched in WGAC's afternoon slot, a new morning show at WGAC began to anchor listeners to the station throughout the day. Harley Drew (WBBQ's former superstar announcer), Mary Liz Nolan and their producer, Matt Stovall, took on the duties as the on-air morning personalities. Both Drew and Nolan had been in music radio for many years before signing on with WGAC. The duo learned how to take the talk radio morning broadcast by storm, and they did so together—with the aid of an occasional shot of a certain candy-coated chocolate. With their obvious chemistry and their passion to alert people about what is going on in the community, Drew and Nolan have been undeniable assets to WGAC.

The twosome took this new journey together and made it their own. The pair, like Rhodes, has succeeded because they know how to relate to the community. With Rhodes in the afternoon and Limbaugh in the middle of the day, WGAC needed something to tie all of the programs

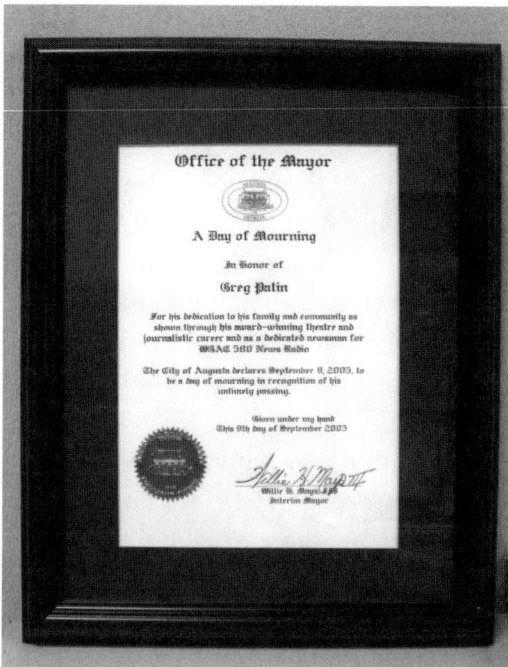

Interim Augusta mayor Willie Mays proclaimed September 9, 2005, to be a day of mourning for WGAC reporter Greg Patin.

together, and Drew and Nolan were just the people for the job. The WGAC morning show consisted of mostly news and only a little discussion, but listeners are loyal to the station's morning block.[266]

Even as WGAC began to establish its morning show as a news-driven product that did its own news gathering, rather than ripping and reading from other media, Nolan knew that the station had to have an on-the-scene reporter—someone who would attend government meetings and investigate tips from listeners or informants. Nolan hired a person for that position who was already associated with WGAC, Greg Patin, who had been news director in the early 1990s. Before long, Patin reestablished his reputation around Augusta as a hard-hitting newsman. He won the Gabby Award from the Georgia Association of Broadcasters for his spot news coverage of the Reinaldo Rivera trial, a serial killer who raped and murdered four young Augusta women. The Radio and Television News Directors Association also gave Patin a Murrow Award for his continuing coverage of the Rivera story. Unfortunately, Patin's final tenure at WGAC was cut short when he died suddenly in September 2005.

The Voice of the Garden City for Seventy Years

After only a few months of looking for a replacement for Patin, Nolan persuaded another former WGAC employee, Scott Hudson, to take Patin's place. Hudson, who was just finishing up college and looking forward to law school, was reluctant to get back into radio. Nolan enticed Hudson with an offer: if he took the job, he could chase any story he wished without editorial interference. In other words, if a story was newsworthy and truthful, Nolan committed to seeing it get on the air.

Having lived most of his life in Augusta, Hudson knew that there were problems in local politics that needed to be revealed, and he came to see the job as a chance to right some wrongs in his hometown. Having no background in journalism but armed with a thorough knowledge of the law, Hudson brought a brand-new style of reporting to the Garden City. When he has had to obtain information under the state Freedom of Information Act, Hudson has been known to quote the law from memory. As a result, he has been successful in obtaining information that other reporters gave up on getting. Being a former disc jockey, Hudson's reports immediately became noticeable for his use of sound effects, music and sometimes humor. Hudson made it clear to the audience from day one that he was an activist journalist, and he intended to tell the truth without leaving out any blemishes that might offend listeners or advertisers.

Hudson broke just such a story in 2011. It involved Columbia County commissioner Scott Dean, who was in an inappropriate relationship with a county employee under his direct control. Georgia law prohibits such relationships. To find out details of the situation, Hudson literally phoned all over the world to find the employee's estranged husband, a Special Forces soldier serving in Iraq. The estranged husband provided cellphone records and GPS tracking information that proved the illicit relationship existed beyond any doubt. Hudson's scoop encouraged others to come forward with allegations that Dean had repeatedly molested one of his adopted daughters. Dean is now serving twenty years in prison.

"Scott's unique style of reporting has led to what I consider a revival of purpose among journalists in the CSRA," says Columbia County chief magistrate judge Bobby Christine. "As a consumer of news, I feel Scott has raised the bar for other reporters to follow, and that revival of purpose for what's known as the fourth estate is healthy for our republic."

WGAC has collected numerous awards for its news programming from the Georgia Association of Broadcasters and the Radio and Television News Directors Association. *Photo by Debbie van Tuyll.*

Another of Hudson's goals has been to bring attention to Richmond County's racial politics. By official charter and racial gerrymandering, Augusta was divided between five white commissioners and five black commissioners when Hudson began his reporting career at WGAC. It was a "Gentleman's Agreement" that if the mayor was white, the mayor pro tem would be black. This arrangement led to questionable uses of power on both sides of the color aisle. Hudson has specialized in covering this type of story. Such a story involved Betty Beard, widow of popular Augusta commissioner Lee Beard, who joined the commission after her husband's death. Hudson discovered that Beard had helped the company Edmondson and Gallagher direct a large charitable gift to the city and used part of it to pay for a city employee's gastric bypass surgery. There were questions at the time as to whether the gift was made in return for Beard's influence with the commission on behalf of Edmondson and Gallagher, which was interested in some redevelopment projects in downtown Augusta. Both Beard and Edmondson and Gallagher denied any wrongdoing.[267]

"There is no question that Scott Hudson and WGAC's reporting has helped change the nature of our government," said Augusta commissioner Joe Jackson. "They have all [as a news team] gone places the other media refused to touch, and each time came out with a scoop that eventually saved taxpayers money." Jackson noted that while a typical Scott Hudson report lasts for a little over a minute, Harley, Austin and Mary Liz are given plenty of airtime to flesh out the stories and expand on what Scott has reported, making news much more interactive especially when it involves complicated issues.

Hudson is a notable persona around Augusta, due in part to his trademark reporter's fedora and double-breasted suit. As Fuqua told Hudson in person, and later wrote about in his memoir, a person's outward appearance reflects the inner self and influences his credibility. Hudson today dresses more like a television reporter than like someone from radio in order to come across as the crusading reporter that he is.

Talk radio pulls in listeners that most other types of radio formats cannot. Talk radio listeners are loyal, and hosts on talk radio become more personable to the listener due to the relationship they have with their audience. Not only do the local hosts give listeners a feeling of community, but the daily connection with the station also makes listeners feel closer to the hosts on a more intimate level.

The very local nature of Rhodes's and Drew and Nolan's shows generates dedicated listeners who want to know what the hosts have to say and who also desire the ability to interact with the personalities. Talk radio gives listeners a sense of involvement that has helped to make the station very successful. WGAC offers local and national information to people with a variety of local and national talk show personalities. They had "real bodies and real personalities doing the job," and that's what set them apart.[268]

When Drew and Nolan started at WGAC, their morning show consisted of three people. After sixteen years of being on the air, their award-winning show had expanded to about a seven-person show. Other on-air personalities include Keith Beckum, their producer; Chad Bennett, who does traffic; Ashley Brown with sports; Scott Hudson, who does investigative reporting; and Steve Smith, a meteorologist.

People who have listened to the *Austin Rhodes Show* or *Augusta's Morning News with Harley Drew and Mary Liz Nolan* from the beginning have heard the programs expand and mature with time and age. There is no station in Augusta today that can compete with WGAC regarding the quality of its radio personalities or the dedication of its listeners. WGAC finally has all the ingredients it needs to be a successful talk radio station.[269]

EPILOGUE

**Abigail Blankenship, Catherine Collingsworth,
Tiffannie Meador and Scott Hudson**

The radio industry has been written off as dead many times. It began as a technology solely for maritime uses. Later, the advent of television had many saying that radio was obsolete. Radio had other plans, though, and met the challenge of the new broadcast medium by reinventing itself as an intensely local medium whose listenership came primarily from people out and about in their cars. Had it not been for obstinate radio executives, AM radio could have long ago gone the way of the telegraph.

This was due to the progress of sound recording technology marching toward producing recordings with greater fidelity. Sound recording technology allowed recorded music to sound almost as good as live music thanks to men like Phil Specter, who developed the "Wall of Sound" technique; Les Paul, who developed the method of "tracking" as opposed to recording a song all at once; and Brian Epstein, who furthered the process by creating the groundbreaking sounds of the Beatles. But AM radio just did not have the ability to reproduce that high-quality sound. This would have sounded the death knell for AM if not for the emergence of three formats that have become standards on the AM band. Those formats—talk radio, religious programming and

Left: Austin Rhodes has been WGAC's controversial afternoon news-talk show host for nearly twenty years. *Photo by Debbie van Tuyll.*

Below: Mary Liz Nolan's quick stash of M&Ms are within reach at the news station, where she is a WGAC on-air news anchor and is a source of dependable news updates in the CSRA. *Photo by Patricia Johnson of Saturated Simplicity Photography.*

sports programming—offered the saving grace for the AM band. Other factors were important, too, in the revitalization of AM radio.

One of those was deregulation in the 1980s. Deregulation allowed for the creation of radio conglomerates like Beasley Broadcasting, WGAC's present owner. Corporations were able to bring economies of scale to the industry that had been impossible when each station in a market had to have a separate and unique owner. The ability to own multiple stations allowed the conglomerates to use profits from performing stations to prop up underperforming AM signals with the hope that one day someone like Rush Limbaugh would come along and reinvigorate the AM's moneymakers again.

Deregulation and talk radio have been crucial in WGAC's survival, but so has the willingness of the station's owners over the years to do whatever they needed to do to stay on the air. From the innovative J.B. Fuqua to the corporate powerhouse Beasley Broadcasting, WGAC's owners have taken their public service responsibilities seriously, and they have always sought better and more practical ways to serve their listeners in and around Georgia's second-largest city.

On August 10, 2011, WGAC took a practical step that astounded many: it added a second frequency to carry its programming, and that frequency was on the FM band. Owners and managers determined that it was finally time to take a plunge into the FM spectrum for WGAC. This was one of the biggest changes yet for WGAC, to broadcast simultaneously on both AM and FM.[270] Sound quality was not the only factor in the decision to add the second frequency—the practicality of FM waves was important as well.[271] About 90 percent of radio listeners never leave the FM band, thus they never have an opportunity to come across WGAC's news-talk programming. By adding an FM broadcast, Beasley Broadcast Augusta opened up an opportunity for radio listeners to find WGAC whenever they hit their seek and scan buttons. The hope was that not only would listeners seek, but they would also find, stop, listen and stay.

WGAC general manager Kent Dunn believes that demographics have a lot to do with why most people today listen to FM more frequently than AM. Research done by station managers "found that [WGAC's]

Left: The late Gregory Patin, recipient of an Edward R. Murrow Award for his excellence in journalism, maintained an energetic personality on the WGAC news team. *Photo by Patricia Johnson of Saturated Simplicity Photography.*

Below: *Left to right*: Tim Lawandus, Denise Warkenthien, Sarah Watson, Joyce Jackson, Kent Dunn and Lillian Smith. WGAC's staff combines thoughts throughout a routine business meeting in the early hours of the day. *Photo by Patricia Johnson of Saturated Simplicity Photography.*

The Voice of Democracy Award was presented to WGAC for its diligence in 1981–82. The Voice of Democracy is awarded by the Veterans of Foreign Wars (VFW). *Photo by Patricia Johnson of Saturated Simplicity Photography.*

George Beasley of Beasley Broadcasting is the present owner of WGAC. *Beasley Broadcasting.*

WGAC operates today from new studios in Evans, Georgia, a suburb of Augusta. *Photo by Debbie van Tuyll.*

demographic and ratings were getting older." The move to FM was to put the station in a position where younger listeners, who never developed the habit of listening to AM, can find it. The station "needed to be on the FM band where sampling occurs," Dunn added.[272]

There were other reasons for the switch, too. Modern building materials like steel and fluorescent lights hamper AM signals more than FM signals. The station's heritage was another reason WGAC has remained committed to the AM band. "The reason why we didn't switch just straight over to FM is two reasons: number one, 580 AM is a heritage frequency here in Augusta, and number two, it goes a whole heck of a lot further than any other AM station in this town," Hudson said.[273]

The most important factor behind the change, though—the one that truly guided the decision—was the station management's commitment to listeners. When those early risers in Augusta are driving to work, they no longer had to switch to the AM dial to hear their local news and sports. They can stay where they likely already are, on the FM dial. Those who already have the AM habit will still be able to find the station where it's been for nearly sixty years. Having an FM presence and being on the original AM frequency will ultimately increase the station's listeners,[274] according to Kent Dunn.

The frequency change implied nothing for programming. WGAC is still committed to providing local programs with local personalities such as Austin Rhodes, Harley Drew, Mary Liz Nolan and Scott Hudson talking about topics that specifically affect the people within the Augusta area. Even with the switch to the FM band, WGAC has always been Augusta's station for local content. Its personalities, like Fisher, Moore, Rhodes, Nolan and Drew, have never been just voices on the radio. They are presences within the community, and they are trusted to report on local stories and institutions.

Some of WGAC's proudest work has resulted in the removal of corrupt politicians from office and uncovering stories that some would have preferred to remain hidden, but the station proudly reports the good things that happen in Augusta, too. Its dedicated reporting of the Masters Golf Tournament each year led it to be named the only radio station with traffic and parking reports from the Augusta National during

Not only is WGAC consistently named one of the best radio stations in Georgia, its listeners also routinely select it for *Augusta Magazine*'s annual Best of Augusta award.

Masters Week. WGAC provides wall-to-wall coverage of the tournament and offers official traffic information from the golf course every five minutes during peak drive times. It is the station even visitors tune in to when they're in Augusta for the tournament.

WGAC has also proudly launched the national careers of people whose first radio experience was in Augusta. Herman Cain, a millionaire businessman whose bid for the nomination in the 2011 Republican presidential primary went down in flames due to allegations of sexual misconduct, got his start toward national ascendency as a fill-in host for Austin Rhodes.

Barry Yon, host of *Voice of the People* in the 1970s (as "Bob Young"), married young computer whiz Kim Komando. She hosts a computer and tech show, calling herself "the Digital Goddess," and WGAC became one of the first stations in the nation to air the program. The *Kim Komando Show* is now the top-rated weekend program across the nation. Recently,

News Director Mary Liz Nolan and General Manager Kent Dunn share a laugh. *Photo by Debbie van Tuyll.*

in a video tribute to Harley Drew while celebrating his fiftieth year in broadcasting, Komando paid tribute to WGAC for taking the chance on an unknown radio host with a very specialized show.

According to *Talk Radio*, a book about the history of talk radio and why it became a popular radio format, "many talk show hosts are as famous for the audiences they offend as for those they entertain and attract."[275] This is certainly true of Austin Rhodes. In his afternoon show, while he does inform, he gives his own commentary on the news he is reporting and does not hold back his opinion, which can entertain as well as offend. However, as Station Manager Dunn observed, "When you have someone talking about what's wrong with the community, you have a chance to make it right." Without an Austin Rhodes talking about the actions of those crooked people, wherever they are, listeners might never know about what they had done and would not be able to put a stop to it, Dunn said.[276] Community focus is the other reason why Rhodes and WGAC have been so successful. The station's local programming deals with issues and topics that listeners are interested in and passionate about.

In the true spirit of talk radio and WGAC tradition, Rhodes will take callers, too, to gain their insights on a topic. Callers not only contribute to content, but they have also helped improve the technology that supports talk radio. Because of talk radio, stations that use this format today are equipped with "phone systems that handle more than 30 incoming lines" and "more advanced versions [that] can create a caller database that can include telephone number, address, regular topic of interest, occupation, birthday and zip code."[277] These technologies helped streamline the call-in process and contribute to the success of talk radio.

Talk radio has spurred the development of other new technologies as well. WGAC was one of the first radio stations to use the power of the Internet to stream local programming. Rhodes also stays in contact with his listeners through social media such as Twitter, and one of the prime features of the station's website is investigative reporter Scott Hudson's blog.

Even with the innovations that WGAC has achieved, including the technological and social innovations, it has always remained a station for the people in the Augusta area. WGAC has received many awards over the years and has remained a recognized radio station in the area. This is because its managers and owners are constantly looking to the future to figure out which new development will best serve listeners. According to Mary Liz Nolan, who has been with WGAC since 1995, the radio station has expanded in a number of ways beyond pure talk radio, such as putting more content on its website for people to view if the radio content is not enough for them. The addition of video news content on the WGAC website allows listeners to put a face to the radio personality, which has been virtually absent in the past.[278]

However, WGAC has still kept its talk radio roots with personalities like Harley Drew and Austin Rhodes, and with each having his own personality and opinions, it has contributed to the radio station's success and given the talk show hosts personalities to make them recognizable. Austin Rhodes, in particular, helped the radio station become more locally oriented.[279]

With more than seventy years of on-air presence, this radio station has broken barriers and set standards for talk radio in ways other stations have not thought of nor would dare to put into action. As much as WGAC has changed, the station's dedication to its listeners and its emphasis today on local content continues to capture the attention of audiences, just as the station has done ever since December 1, 1940.

NOTES

PREFACE

1. Guglielmo Marconi, "Wireless Telegraphic Communication," *Nobel Lectures: Physics, 1901–1921* (Singapore: World Scientific Publishing Company, 1998), 196–222.
2. Christopher H. Sterling and John Michael Kittross, *Stay Tuned: A History of American Broadcasting*, 3rd ed. (New York: Taylor and Francis, 2002), 43–44.
3. Ibid.
4. Encyclopedia Titanica, "Mr. Harold McBride," http://www.encyclopedia-titanica.org/titanic-survivor/harold-sydney-bride.html.

CHAPTER 1

5. Don O'Briant, "J.B. Fuqua (1918–2006)," New Georgia Encyclopedia, http://www.georgiaencyclopedia.org/nge/Article.jsp?path=/BusinessIndustry/TopCompaniesEntrepreneursBusines&id=h-1914.
6. Ibid.

7. Ibid.

8. Ibid.

9. Ibid.

10. Ibid.

11. Ibid.; *Augusta Chronicle*, "WRDW Will Give Salute to WGAC," December 1, 1940.

12. Scott Hudson interview, November 21, 2011; *Life*, "Local Editor," April 27, 1942, 77.

13. Our Georgia History, "Georgia History Timeline," ourgeorgiahistory. com/year; Scott Hudson interview; Augusta and Environs Picture Post Cards in Color, "John P. King and Sibley Cotton Mills, Augusta Ga.," as presented in the Digital Library of Georgia, http://dlg.galileo.usg. edu/meta/html/gaec/hagp/meta_gaec_hagp_aep002.html.

14. Scott Hudson interview, November 21, 2011.

15. *Augusta Chronicle*, "WGAC Asks Permit for Lower Wave," December 3, 1944; Scott Hudson interview, November 21, 2011.

16. *Augusta Chronicle*, "WRDW Will Give Salute to WGAC"; *Augusta Chronicle*, "Programs Begin on Station WGAC," December 2, 1940.

17. *Augusta Chronicle*, "Programs Begin on Station WGAC."

18. Ibid.

19. Michele Hilmes, *NBC: America's Network* (Berkeley: University of California Press, 2007), 91–97.

20. *Augusta Chronicle*, "WRDW Will Give Salute to WGAC."

21. *Augusta Chronicle*, WGAC ad, December 1, 1940.

22. Jim Davis phone interview, October 28, 2011; O'Briant, "J.B. Fuqua"; Scott Hudson interview, November 21, 2011.

23. *Augusta Chronicle*, "WGAC to Broadcast Noon Stock Prices," January 6, 1941.

24. Scott Hudson interview, September 19, 2011.

25. Jim Davis interview, October 28, 2011.

26. *Augusta Chronicle*, "WGAC Asks Permit for Lower Wave."

27. *Augusta Chronicle*, "Sale of Station WTNT Is Given FCC Approval," September 17, 1949.

28. *Augusta Chronicle*, "Video Set-up in Augusta Described to Lions Club," December 5, 1953.

29. Jim Davis interview, October 28, 2011.

30. Ibid.

31. Martha Nolan McKenzie, "Executive Life: Overcoming His Torment from Within," *New York Times*, March 31, 2002; Scott Hudson interview, November 21, 2011.

32. O'Briant, "J.B. Fuqua."

33. J.B. Fuqua, *Fuqua, A Memoir: How I Made My Fortune Using Other People's Money* (Atlanta, GA: Longstreet Press, 2002).

34. O'Briant, "J.B. Fuqua"; McKenzie, "Executive Life"; Rhonda Rowland, "CNN Health," CNN, http://articles.cnn.com/2002-07-16/health/cov.depression.ceo_1_fuqua-depression-stress?_s=PM:HEALTH.

35. O'Briant, "J.B. Fuqua."

CHAPTER 2

36. *Augusta Chronicle*, "War Relief Drive to Start Monday," October 20, 1940.

37. John C. Davenport, *The Attack on Pearl Harbor: The United States Enters World War II* (New York: Chelsea House, 2009), 1–7.

38. *Life*, "Radio: War Ends Rich Era and Brings Fresh Problems," April 27, 1942, 70.

39. Ibid., 70, 72.

40. *Life*, "WGAC: Small Station Mirrors Huge Industry," April 27, 1942, 76.

41. William Bates Williams, "John Shaw Billings: The Man Behind the Editor" (master's thesis, University of South Carolina, 1975), 3.

42. *Augusta Chronicle*, "Personals," April 23, 1946.

43. *Augusta Chronicle*, "Historical Society to Hold June Session," May 16, 1947.

44. *Augusta Chronicle*, "Augusta Station Shown in Life," April 24, 1942; Williams, "John Shaw Billings," 61.

45. *Augusta Chronicle*, "Augusta Station Shown in Life."

46. *Life*, "Sundays," "Local College Girl," photo cutlines, April 27, 1942, 76.

47. *Life*, WGAC pictorial, April 27, 1942, 76–77.

48. Ibid., 77.

49. Mark Dunn, "Fort Gordon," New Georgia Encyclopedia, June 10, 2005, http://www.georgiaencyclopedia.org/nge/Article.jsp?id=h-1321.

50. *Augusta Chronicle*, radio listings, June 30 and July 1, 1942.

51. *Augusta Chronicle*, radio listings, June 29, 1942.

52. *Augusta Chronicle*, radio listings, July 1, 1941.

53. *Augusta Chronicle*, "FDR to Make Major Speech Tonight," May 27, 1941.

54. Ibid.

55. Associated Press, "British Mop Up Bengasi Sector," in *Augusta Chronicle*, February 9, 1941.

56. Fuqua, *Fuqua*, 29–30.

57. *Augusta Chronicle*, "Volunteer Workers to Launch Red Cross Role Call Today," November 11, 1941.

58. Ibid.

59. *Augusta Chronicle*, "Nursing Leader to Visit Here," March 8, 1941.

60. *Augusta Chronicle*, "Three Division Chairmen Named for War Bond Drive," January 5, 1944; *Augusta Chronicle*, "Actual Scenes of War Are on Display in City," February 3, 1944.

61. Michael Socolow, "News Is a Weapon: Domestic Radio Propaganda and Broadcast Journalism in America, 1939–1944," *American Journalism* 24, no. 3 (2007): 117.

62. Ramon Girona and Jordi Xifra, "The Office of Facts and Figures: Archibald MacLeish and the 'Strategy of Truth,'" *Public Relations Review* 35 (2009): 287–90.

63. Michael S. Sweeney, *The Military and the Press: An Uneasy Truce* (Evanston, IL: Northwestern University Press, 2006), 86.

64. Gerd Horton, *Radio Goes to War: The Cultural Politics of Propaganda During World War* II (Berkeley: University of California Press, 2002), 45; James Spiller, "This Is War! Network Radio and World War II Propaganda in America," *Journal of Radio Studies* 11, no. 1 (November 1, 2004): 56, http://web.ebscohost.com/ehost/pdfviewer/pdfviewer?sid=0eadf0e2-12fa-498b-8df8-fcc0849b16d7%40sessionmgr104&vid=7&hid=125.

65. Spiller, "This Is War!" 57.

66. Ibid., 56.

67. Ibid., 61.

68. Horton, *Radio Goes to War*, 46–47.

69. Spiller, "This Is War!" x.

70. *Augusta Chronicle*, "Reds Smash Stalingrad Attacks," September 30, 1942.

71. *Augusta Chronicle*, "Soviet Juggernaut Pounding Nearer to Rostov: Reds Overrun Bitter Nazi Resistance," January 9, 1943.

72. *Augusta Chronicle*, "Germans Slaughter All Men in Village: Buildings Are Razed and Name of Town Erased in Nazi Bloodbath," June 11, 1942.

73. Donald G. Godfrey, "Radio News," *History of the Mass Media in the United States: An Encyclopedia* (New York: Routledge, 1998): 567, http://web.ebscohost.com/ehost/pdfviewer/pdfviewer?sid=0eadf0e2-12fa-498b-8df8-fcc0849b16d7%40sessionmgr104&vid=9&hid=125.

74. Ibid., 568.

75. *Augusta Chronicle*, "USO Broadcast Here Tonight to Hit Network," October 13, 1942.

76. John Dunning, "George Hicks," *The Encyclopedia of Old-Time Radio* (New York: Oxford University Press, 1998), 498.

77. Old Time Radio, "Victory in Japan Day," http://www.otr.com/vj.html.

78. Ibid.

79. Jennifer Brooks, "Winning the Peace: Georgia Veterans and the Struggle to Define the Political Legacy of World War II," *The Journal of Southern History*, 66, no. 3 (2000): 556.

80. *Augusta Chronicle*, "Newspapers Helped End Long Rule of Corrupt Cracker Party," August 29, 2010, http://chronicle.augusta.com/225/2010-08-29/newspapers-helped-end-long-rule-corrupt-cracker-party.

81. Ibid.

82. Ibid.

Chapter 3

83. *Augusta Chronicle*, "WGAC Asks Permit for Lower Wave."

84. *Augusta Chronicle*, "Don't Miss It," WGAC ad, July 19, 1946.

85. Mary Carter Winter, "Numerous Changes Are Noted in the Business District of the City," *Augusta Chronicle*, July 28, 1946.

86. *Augusta Chronicle*, "Founder of Radio, TV Outlets Is Honored," November 9, 2002.

87. *Augusta Chronicle*, "WGAC—Augusta's Home Owned Station," February 24, 1946.

88. *Augusta Chronicle*, "Work Is Begun on WGAC Studio," February 24, 1946.

89. *Augusta Chronicle*, "WGAC Goes Over to 58 on Dial at 5 o'clock This Afternoon," December 1, 1946.

90. National Film and Sound Archive of Australia, "Effects of TV on Radio," January 25, 2012, http://dl.nfsa.gov.au/module/104.

91. Jason Mittell, "Before the Scandals: The Radio Precedents of the Quiz Show Genre," in Michele Hilmes and Jason Loviglio, *Radio Reader: Essays in the Cultural History of Radio* (New York: Routledge, 2002), 319–22; J. Fred MacDonald, *Don't Touch That Dial! Radio Programming in American Life, 1920–1960* (web book), http://jfredmacdonald.com/metamorphosis.htm.

92. MSNBC, "About 'Meet the Press,'" http://www.msnbc.msn.com/id/3403008/ns/meet_the_press-about_us/t/about-meet-press/#.TyakFRzIcRg, January 30, 2012; MacDonald, *Don't Touch That Dial!*

93. MacDonald, *Don't Touch That Dial!*

94. Susan L. Brinson, *The Red Scare, Politics, and the Federal Communications Commission, 1941–1960* (Westport, CT: Greenwood Publishing, 2004), 131–32; F. Leslie Smith, John W. Wright II and David H. Ostroff, *Perspectives on Radio and Television: Telecommunications in the United States* (Mahwah, NJ: Lawrence Erlbaum Associates, Inc., 1998), 59; Brian Fitzgerald, *McCarthyism: The Red Scare* (Minneapolis, MN: Compass Point Books, 2007), 16; Oral History Interview with Clifford Durr, December 29, 1974, Interview B-0017, Southern Oral History Program Collection, Documenting the American

South website, http://docsouth.unc.edu/sohp/B-0017/excerpts/excerpt_8767.html.

95. MacDonald, *Don't Touch That Dial!*

96. Irving Settel, *A Pictorial History of Radio* (New York: Crosset and Dunlap, 1967).

97. *Augusta Chronicle*, radio listings, September 25, 1941, and September 27, 1948.

98. Esther Young, "Intricate Plans Necessary for Sammy Kaye Broadcast," *Augusta Chronicle,* January 13, 1946.

99. Young, "Intricate Plans Necessary."

100. William A. Richter, *Radio: A Complete Guide to the Industry* (New York: Peter Lang Publishing, Inc., 2006).

101. *Augusta Chronicle*, WGAC ad, May 31, 1946.

102. *Augusta Chronicle*, "Attend 'Breakfast in Augusta,'" photo headline, June 6, 1946; *Augusta Chronicle*, "Perrin School Sponsoring Party," April 6, 1947.

103. *Augusta Chronicle*, "Gospel Group 'Walked in the Light' for Half a Century," December 29,1990.

104. Don Rhodes, *Say It Loud! My Memories of James Brown, Soul Brother No. 1* (Guilford, CN: Lyons Press, 2009).

105. Scott Hudson interview, October 1, 2011.

106. Ibid.

107. Clint Engel, "Fuqua Pays Homage to His Business Roots," *Augusta Chronicle*, November 6, 1990.

108. Harley Drew interview, October 2011.

109. *Augusta Chronicle*, "Radio Station, WBBQ, to Begin Broadcasting This Morning," January 12, 1947.

110. Ibid.

111. *Augusta Chronicle*, "Claiming Right to Broadcast All Out-of-Town Ball Games, WGAC Operators Sue WBBQ," February 11, 1949.

112. Ibid.

113. *Augusta Chronicle*, "Augusta Radio Suit Argued; Will Be Resumed on Friday," February 26, 1949.

114. Ibid.

115. *Augusta Chronicle*, "Suit Settled as WBBQ Pays Damages, Gets Rights on Out-of-Town Games," March 11, 1949.

116. *Augusta Chronicle*, "Claiming Right to Broadcast."

117. *Augusta Chronicle*, "Handsome Harley Drew Signs Off," July 23, 1989.

118. Ibid.

119. Clint Engel, "Long Reign at Top Ends for WBBQ," *Augusta Chronicle*, February 3, 1990.

120. Online Radio Stations, "Urban Music," http://www.onlineradiostations.com/music-genre/urban-music/urban-music-styles-include-rap-hip-hop-and-urban-pop.html.

121. Engel, "Long Reign at Top Ends for WBBQ."

122. Ibid.

123. Clint Engel, "Monthly Radio Ratings Report Shows WBBQ Back on Top," *Augusta Chronicle*, February 24, 1990.

124. Federal Communications Commission, "FCC Revises National Multiple Radio Ownership Rule and Local Ownership Rule in Accordance with the Telecommunications Act of 1996," http://transition.fcc.gov/Bureaus/Mass_Media/News_Releases/1996/nrmm6009.txt; Federal Communications Commission, "Rules Adopted in the Quadrennial Review Order," http://transition.fcc.gov/ownership/rules.html.

125. Mike Wynn, "Stations Change Tunes," *Augusta Chronicle*, February 14, 1993.

126. Wynn, "Stations Change Tunes."

127. Harley Drew interview.

128. Chris Cady, "Remembers 'Gentleman' of WBBQ," *Augusta Chronicle*, December 29, 1997.

129. *Augusta Chronicle*, "The Changing Face of Augusta Radio," February 4, 2002.

130. Henry Greene, "Station May Swap Rock for Real Floating Jazz," *Augusta Chronicle*, May 7, 1973.

131. Clint Engel, "WBBQ Falls from No. 1 Spot for the First Time since 1962," *Augusta Chronicle*, February 3, 1990.

132. Harley Drew interview.

133. Cady, "Remembers 'Gentleman' of WBBQ."

134. Ibid.

135. *Augusta Chronicle*, "Mourns Loss of 'WBBQ Mobile News,'" June 14, 1998.

136. Kent Dunn interview, November 4, 2011.

137. *Augusta Chronicle*, "Changing Face of Augusta Radio."

138. Kent Dunn interview, November 4, 2011.

CHAPTER 4

139. Andy Bennett, *Cultures of Popular Music* (Buckingham, England: Open University Press, 2001), 8–10; Roger Chapman, ed., "Rock and Roll," *Culture Wars: An Encyclopedia of Issues, Viewpoints, and Voices*, vol. 1 (Armonk, NY: M.E. Sharp, Inc., 2010), 475.

140. Scott Hudson interview, November 4, 2011.

141. Cindy O'Brien, Carol Spires and Connie Bowles interview, October 13, 2011.

142. Keith Beckum interview, Feburary 3, 2012.

143. O'Brien interview, November 11, 2011.

144. O'Brien, Spires and Bowles interview.

145. Scott Hudson interview, November 4, 2011.

146. Ibid.

147. O'Brien, Spires and Bowles interview.

148. George Fisher interview (written question-and-answer interview), in possession of O'Brien, Spires and Bowles.

149. Saatchi Gallery Contemporary Art, Education Department, "1950: The Beginnings of Rock and Roll," http://esto.es/rock/english/history.htm; Georgia Radio Museum and Hall of Fame, "George Weiss," http://www.grhof.com/GEORGE%20WEISS.htm.

150. Peter Guralnick, *Last Train to Memphis: The Rise of Elvis Presley* (New York: Little, Brown and Company, 1994) , 200–219.

151. Kent Dunn interview, November 21, 2011.

152. David E. Scherman, "Transistor Craze—There's No Escape," *Life*, November 24, 1961, 23.

153. David Crowley and Paul Heyer, eds., *Communication in History: Technology, Culture, Society*, 5th ed. (Boston: Allyn & Bacon, 2007), 234.

154. Ibid.

155. The Inflation Calculator, http://www.westegg.com/inflation.

156. Harley Drew interview by Scott Hudson, October 2011.

157. Harley Drew interview.

158. *Augusta Chronicle*, "Local Station to Become CBS Affiliate Monday," February 28, 1965, 2.

159. *Augusta Chronicle*, WGAC ad, February 28, 1965, 5.

160. Harley Drew interview.

161. Ibid.

162. Peter Fortale and Joshua E. Mills, "Radio in the Television Age," *Communication in History: Technology, Culture, Society*, ed. David Crowley and Paul Heyer (New York: Allyn and Bacon, 2006), 230–31.

163. Ibid.

164. Ibid.

165. WRDW-TV, "Special Assignment: James Brown Helps Augusta Get Back on the Good Foot After Race Riots," posted May 12, 2010, http://www.wrdw.com/home/headlines/93532289.html.

166. Edward J. Cashin, *The Story of Augusta* (Spartanburg, SC: Reprint Co., 1991), 299; Paine College, "Paine College's Class of 1970 Returns Forty Years Later to Participate in Commencement Exercises," posted April 27, 2011, http://www.paine.edu/blog/post/2011/04/27/Paine-Collegee28099s-Class-of-1970-Returns-Forty-Years-Later-to-Participate-in-Commencement-Exercises.aspx.

167. Paine College, "Paine College's Class of 1970 Returns."

168. Ibid.

169. Ibid., 302; WRDW-TV, "Former Councilman Remembers 1970 Augusta Race Riot," posted June 4, 2010, http://www2.wjbf.com/news/2010/may/12/former_councilman_remembers_1970_augusta_race_riot-ar-287383.

170. Edward Cashin, *An Informal History of Augusta* (Augusta, GA: Richmond County Board of Education, n.d.), 100–105.

171. Ibid., 104.

172. Ibid.

173. Ibid.

174. Jim Miller, "Augustan Indicted in Triple Slaying," *Augusta Chronicle*, July 24, 1970.
175. Cashin, *Informal History*, 234.
176. "George Fisher Interview."

CHAPTER 5

177. National Criminal Justice Reference Service, "America's Drug Abuse Profile," https://www.ncjrs.gov/htm/chapter2.htm.
178. "George Fisher Interview," in the personal collection of Cindy O'Brien, Augusta, Georgia, obtained October 13, 2011.
179. "George Fisher Interview."
180. Seth Taylor, "Missing You," article in the George Fisher papers in the possession of Cindy O'Brien, Augusta, Georgia.
181. Ibid.
182. Bob Young, "Closeup," *Richmond County (Augusta) News*, July 4, 1978; e-mail from Cindy O'Brien, February 5, 2012, in possession of the editors.
183. "WBIA's Biography of George Fisher," from the personal collection of Cindy O'Brien, Augusta, Georgia, obtained October 13, 2011.
184. "WBIA's Biography of George Fisher."
185. "George Fisher Georgia Hall of Fame Induction Keynote," lecture, October 4, 2008, from the personal collection of Cindy O'Brien, Augusta, Georgia, obtained October 13, 2011.
186. O'Brien interview.
187. Ibid.
188. Scott Hudson interview, Augusta, Georgia, November 3, 2011.
189. O'Brien interview.
190. O'Brien, Spires and Bowles interview.
191. Ibid.
192. Taylor, "Missing You."
193. George Fisher interview (written question-and-answer interview), in possession of O'Brien, Spires and Bowles.

194. O'Brien, Spires and Bowles interview.

195. O'Brien, Spires and Bowles interview.

196. Joe Frank Harris, letter to George Fisher, July 11, 1988.

197. Taylor, "Missing You."

198. Ibid.

199. Ibid.

200. "WBIA's Biography of George Fisher."

201. "Awards and Community Awards," from the personal collection of Cindy O'Brien, Augusta, Georgia, October 13, 2011).

202. *Augusta Chronicle*, "Rotary Club Gives Media Awards," March 17, 1984.

203. O'Brien, Spires and Bowles interview.

204. Ibid.

205. O'Brien interview.

206. Ibid.

207. Scott Hudson interview, November 3, 2011.

208. Ibid.

209. Ibid.

210. Elizabeth Mulherin interview, November 28, 2011.

211. Ibid.

212. Ibid.

213. Dudley Martin, "Mulherin Making Name on the Airwaves, Fields," *Augusta Chronicle*, May 11, 1986.

214. Elizabeth Mulherin interview.

215. Augusta State 1989 Basketball Media Guide, from the personal collection of Elizabeth Mulherin, Augusta, Georgia, November 28, 2011.

216. Elizabeth Mulherin interview.

217. Bernard Mulherin interview.

218. Photos from the personal collection of Elizabeth Mulherin, Augusta, Georgia, November 28, 2011.

219. Cindy O'Brien interview, Art Factory, Augusta, Georgia, October 11, 2011.

220. Bernard Mulherin interview, Augusta, Georgia, November 21, 2011.

221. "Round Town: Front Cover," article from unknown magazine published on December 4, 1954, in the personal collection of Cindy O'Brien, Augusta, Georgia.
222. "WBIA's Biography of George Fisher."
223. Keith Beckum interview, Feburary 3, 2012.
224. Bernard Mulherin interview.
225. Kent Dunn interview, November 21, 2011.
226. Ibid.
227. "George Fisher Interview."
228. "George Fisher Job Description," from the personal collection of Cindy O'Brien, Augusta. Georgia, October 13, 2011.
229. *Augusta Chronicle*, George Fisher obituary, July 10, 1995.
230. Tharon Giddens, "Radio Loses a Legend," *Augusta Chronicle*, July 10 1995; e-mail from Cindy O'Brien, February 5, 2012, in possession of the editors.

Chapter 6

231. Michael C. Keith, *The Radio Station: Broadcast, Satellite, and Internet*, 8th ed. (Burlington, MA: Focal Press, 2010), 11.
232. Roger L. Sadler, *Electronic Media Law* (Thousand Oaks, CA: Sage Publications, Inc., 2005), 67.
233. Sadler, *Electronic Media Law*, 68.
234. Michael C. Keith, *Talking Radio: An Oral History of American Radio in the Television Age* (Armonk, NY: M.E. Sharpe, 2000), 78; Sadler, *Electronic Media Law*, 68.
235. Scott Hudson interview, September 2011.
236. Harley Drew interview, October 2011.
237. Ibid.
238. Keith, *Talking Radio*, 78.
239. Ibid.; Scott Hudson interview, September 2011.
240. David Barker and Kathleen Knight, "Political Talk Radio and Public Opinion," *Public Opinion Quarterly* 64, no. 2 (2000): 149–70.
241. Keith, *Talking Radio*, 79.

242. Kent Dunn interview, November 2011.

243. Beasley Broadcasting 2000 Annual Report filing for the Securities and Exchange Commission, http://www.getfilings.com/o0001021408-01-001243.html, February 5, 2011; Scott Hudson interview.

244. Ibid.

245. I. Yanovitzky and J.N. Cappella, "Effect of Call-In Political Talk Radio Shows on Their Audiences: Evidence from a Multi-Wave Panel Analysis," *International Journal of Public Opinion Research* 13, no. 4 (January 2001): 377–97.

246. Harley Drew interview, November 2011.

247. Austin Rhodes interview, October 10, 2011.

248. Austin Rhodes interview.

249. Ibid.

250. Ibid.

251. Ibid.

252. Ibid.

253. Ibid.

254. Scott Hudson interview, Augusta, Georgia, September 2011.

255. Wilfred M. McClay, "How to Understand Rush Limbaugh," *Commentary* 131, no. 2 (2011): 19.

256. Austin Rhodes interview.

257. Ibid.

258. Mary Liz Nolan interview, October 13, 2011.

259. Austin Rhodes interview.

260. Ibid.

261. Ibid.

262. Ibid.

263. Ibid.

264. Ibid.

265. Ibid.

266. Mary Liz Nolan interview, September 2011.

267. Johnny Edwards, "Questions Surround Beard Donation," Augusta Chronicle, September 19, 2009.

268. Harley Drew interview.

269. Mary Liz Nolan interview, September 2011.

Epilogue

270. *Augusta Chronicle*, "Beasley Broadcast Augusta Changes News/Talk FM Simulcast Station," August 10, 2011.

271. Kent Dunn interview, August 24, 2011.

272. Kent Dunn interview, November 15, 2011.

273. Scott Hudson interview, October 10, 2011.

274. Kent Dunn interview, August 22, 2011.

275. Sandra Ellis and Ed Shane, "Talk Radio," *Museum of Broadcast Communications Encyclopedia of Radio*, ed. Christopher H. Sterling and Michael Keith (New York: Routledge, 2004).

276. Kent Dunn interview, August 22, 2011.

277. Ellis and Shane, "Talk Radio."

278. Mary Liz Nolan interview.

279. Austin Rhodes interview.

ABOUT THE EDITORS

D ebra Reddin van Tuyll is professor of communications at Augusta State University, where she teaches journalism and public relations. She is the author or editor of three other books: *The Civil War and the Press*, *The Southern Press in the Civil War* and *Knights of the Quill: Confederate Correspondents and Their Civil War Reporting*. *Knights of the Quill* was a finalist for the Association for Education in Journalism and Mass Communications' 2011 Tankard Award for the best book in mass communications. Her fourth book, *The Third Stripe: A Cultural History of the Confederate Press*, will be published in 2012. Van Tuyll is married to Dr. Hubert van Tuyll, chair of the ASU Department of History, Anthropology and Philosophy, and they have one daughter, Laura.

S cott Hudson is WGAC's award-winning investigative reporter. He is an honors graduate from Augusta State with a degree in political science and history. He studied media law and ethics with Dr. Van Tuyll at Augusta State. Scott's style of reporting has been noted as being different from others in that not only does Scott consider himself an activist journalist, he also tends to add humor to his reports. In his own words, "I think I am the illegitimate child of Ida Tarbell and Jon Stewart!" Most importantly of all, he is Emerson's daddy.